Smart Innovation, Systems and Technologies

Volume 299

The Smart Innovation, Systems and Technologies book series encompasses the topics of knowledge, intelligence, innovation and sustainability. The aim of the series is to make available a platform for the publication of books on all aspects of single and multi-disciplinary research on these themes in order to make the latest results available in a readily-accessible form. Volumes on interdisciplinary research combining two or more of these areas is particularly sought.

The series covers systems and paradigms that employ knowledge and intelligence in a broad sense. Its scope is systems having embedded knowledge and intelligence, which may be applied to the solution of world problems in industry, the environment and the community. It also focusses on the knowledge-transfer methodologies and innovation strategies employed to make this happen effectively. The combination of intelligent systems tools and a broad range of applications introduces a need for a synergy of disciplines from science, technology, business and the humanities. The series will include conference proceedings, edited collections, monographs, handbooks, reference books, and other relevant types of book in areas of science and technology where smart systems and technologies can offer innovative solutions.

High quality content is an essential feature for all book proposals accepted for the series. It is expected that editors of all accepted volumes will ensure that contributions are subjected to an appropriate level of reviewing process and adhere to KES quality principles.

Indexed by SCOPUS, EI Compendex, INSPEC, WTI Frankfurt eG, zbMATH, Japanese Science and Technology Agency (JST), SCImago, DBLP.

All books published in the series are submitted for consideration in Web of Science.

More information about this series at https://link.springer.com/bookseries/8767

Lakhmi C. Jain · Roumen Kountchev · Kun Zhang ·
Roumiana Kountcheva
Editors

Advances in Wireless Communications and Applications

Smart Wireless Communications: Algorithms and Network Technologies, Proceedings of 5th ICWCA 2021

Springer

Editors
Lakhmi C. Jain
KES International
Shoreham-by-Sea, UK

Roumen Kountchev
Technical University of Sofia
Sofia, Bulgaria

Kun Zhang
Hainan Tropical Ocean University
Sanya, Hainan, China

Roumiana Kountcheva
TK Engineering
Sofia, Bulgaria

ISSN 2190-3018 ISSN 2190-3026 (electronic)
Smart Innovation, Systems and Technologies
ISBN 978-981-19-2257-2 ISBN 978-981-19-2255-8 (eBook)
https://doi.org/10.1007/978-981-19-2255-8

This Springer imprint is published by the registered company Springer Nature Singapore Pte Ltd.
The registered company address is: 152 Beach Road, #21-01/04 Gateway East, Singapore 189721, Singapore

Preface

This book contains the high-quality peer-reviewed research papers presented at the Fifth International Conference on Wireless Communications and Applications (ICWCA 2021) which was held in Hainan, China, during December 17–19. The volume is focused on the presentation of the newest trends and achievements in the development of intelligent algorithms and network technologies in smart communications, with application in the contemporary communication in cellular networks, the physical layer security of cooperative NOMA systems, analysis of the mobile multipath channels, the creation of multilevel coding scheme with polar codes in 5G, building the up-to-date security systems for network intrusion detection, intelligent traffic systems, analysis and enhancement of digital holographic images, IoT-based marine surface communications as well as state-of-the-art real-time precise location technologies, array signal processing, and many others.

The aim of the book is to present the latest achievements of the authors to a wide range of readers: IT specialists, engineers, physicians, Ph.D. students, and other specialists in the area.

Shoreham-by-Sea, UK — Lakhmi C. Jain
Sofia, Bulgaria — Roumen Kountchev
Sanya, China — Kun Zhang
Sofia, Bulgaria — Roumiana Kountcheva
February 2021

Acknowledgments The book editors express their special thanks to chapter reviewers for their efforts and goodwill to help for the successful preparation of the book. Special thanks for Honorary Chair of ICWCA'21 International conference, Prof. Lakhmi C. Jain, General Chairs Prof. Dr. Chong Shen, Prof. Dr. Roumen Kountchev, and Prof. Dr. Kun Zhang, and Program Chair, Prof. Dr. Srikanta Patnaik.

The editors express their warmest thanks to the excellent Springer team which made this book possible.

Contents

About the Editors

Lakhmi C. Jain Ph.D., Dr. H. C., M.E., B.E. (Hons), Fellow (Engineers Australia), is with the Liverpool Hope University. He was formerly with the University of Technology Sydney, the University of Canberra, and Bournemouth University, UK. He founded the KES International for providing a professional community the opportunities for publications, knowledge exchange, cooperation, and teaming. Involving around 5000 researchers drawn from universities and companies worldwide, KES facilitates international cooperation and generates synergy in teaching and research. KES regularly provides networking opportunities for professional community through one of the largest conferences of its kind in the area of KES.

Roumen Kountchev is with the Faculty of Telecommunications, Department of Radio Communications and Video Technologies—Technical University of Sofia, Bulgaria. He has 434 papers published in magazines and conference proceedings and 22 patents. He is Member of Euro Mediterranean Academy of Arts and Sciences; President of Bulgarian Association for Pattern Recognition; Editorial Board Member of IJBST Journal Group; Editorial Board Member of: *International Journal of Reasoning-based Intelligent Systems* and *International Journal of Broad Research in Artificial Intelligence and Neuroscience*; Editor of books for Springer SIST series.

Kun Zhang Ph.D., is Professor and Master's Supervisor. He is Professor of Information and Communication Engineering at Hainan Tropical Ocean University. He is Visiting Professor of the State Key Laboratory of Marine Resources Utilization in the South China Sea, Hainan University; Visiting Master's Tutor of Tianjin University of Science and Technology; Hainan Province High-level Talent, Hainan Province "Top-notch Talent," "South China Sea Master" Youth Project Talent; Hainan Province "515" Project No. 3 Hierarchical Candidates; Senior Member of China Computer Federation (CCF) and Chinese Institute of Electronics (CIE); Member of ACM and IEEE; Scientific and Technological Expert of Hainan Province and Guangdong Province and has published more than 170 papers on EI\CPCI.

Roumiana Kountcheva is Vice President of TK Engineering. She got her Ph.D. at the Technical University of Sofia and became Senior Researcher in 1993. She has 187 publications and five patents and presented 21 plenary speeches at international conferences. She is Member of IRIEM, IDSAI, IJBST Journal Group, and Bulgarian Association for Pattern Recognition. She is Reviewer of WSEAS conferences and journals and Editor of books for Springer SIST series.

Joint Power and Channel Allocation for D2D Communication in Cellular Networks

Linsen Yi, Yucheng He, Yu Zhang, Yan Zhang, and Lin Zhou

Abstract Device-to-device (D2D) communication is a technology that allows devices communicating with other devices directly instead of going through the base station (BS), and it can reduce the burden of base stations and increase the system capacity of cellular networks. In this paper, we raise and analyze a system of D2D communication in cellular networks, which includes a base station in the center and several cellular user equipments (CUEs) coexisting with D2D user equipments (DUEs). Considering the high rate for CUE and low outage probability for DUE, we propose a scheme of resource allocation to improve the ergodic sum rate of CUEs under the constraint to guarantee the outage preference of DUEs. The problem of resource allocation is a non-convex problem, which is usually mathematically intractable. We divided the optimal problem into two sub-questions, power allocation and channel allocation. We firstly maximize the ergodic rate of CUEs under the constraint of reliability of the DUEs to find the optimal power distribution, then a maximum weight bipartite matching is used to find the optimal channel allocation by the Hungarian method. Simulation results demonstrate that the scheme of resource allocation can achieve the desired performance of the system.

1 Introduction

To meet the increasingly high rate requirements of mobile communications, new technologies such as microcell networks and heterogeneous networks have been proposed, which can achieve the high capacity of the culler network by reducing the cell size. However, these technologies still rely on centralized networks and require

L. Yi · Y. He (✉) · Y. Zhang · Y. Zhang · L. Zhou
Xiamen Key Laboratory of Mobile Multimedia Communications, National Huaqiao University, Xiamen 361021, Fujian, China
e-mail: yucheng.he@hqu.edu.cn

Y. He · L. Zhou
State Key Laboratory of Integrated Services Network, Xidian University, Xi'an 710071, Shaanxi, China

L. C. Jain et al. (eds.), *Advances in Wireless Communications and Applications*,
Smart Innovation, Systems and Technologies 299,
https://doi.org/10.1007/978-981-19-2255-8_1

user equipments (UEs) to communicate with a base station (BS). This centralized network is essentially limited by the service capability of the BS, and the access of a large number of UE may cause congestion of the entire network [1]. To alleviate this situation, the concept of device-to-device (D2D) communication has been proposed in recent years. D2D communication allows local peer-to-peer transmission between UE, which can reduce the delay of both parties in communication, reduce the burden on the base station and increase the system capacity of the network [2]. Channel-sharing methods for D2D communications are classified into two parts, overlay D2D communications and underlay D2D communications. In the mode of overlay D2D communications, the D2D user equipments (DUEs) occupy the vacant channels of cellular user equipments (CUEs) for communication. This approach can eliminate interference by dividing channels into two parts, but it is inefficient in terms of channel reuse. In the scheme of underlay D2D communications, multiple DUEs are allowed to work as an underlay with CUEs. It will improve the channel efficiency, but more interference will be brought to the original cellular networks [3].

Recently, there are more and more researches on D2D communication resource allocation to reduce interference. Lee et al. [4] uses stochastic geometry to analyze two different power distribution schemes, centralized power distribution and distributed power distribution, while ensuring the sufficient coverage probability of CUEs with the acceptable interference created by DUEs. The authors of [5] maximized the system capacity while ensuring the performance of CUEs and DUEs. Radaydeh et al. [6] compares the respective benefits of D2D communication under the three resource sharing modes of orthogonal mode, non-orthogonal mode, and cellular mode. The non-orthogonal resource sharing mode has the highest spectrum efficiency, but it is also the most difficult to deal with and it causes more interference to original network.

2 System Model

We consider a system of D2D communication in the single-cell cellular network as shown in Fig. 1. CUEs and DUEs are located around the BS, and each DUE contains a transmitter (DUE_T) and a receiver (DUE_R). Each CUE is allocated a channel by the BS.

In this system model, CUEs are represented by the set $C = \{1, 2, \ldots, I\}$, and DUEs are represented by the set $D = \{1, 2, \ldots, J\}$. In order to reduce interference between channels, it is assumed that the channels are orthogonal, each DUE can only access one channel and each channel can only be reused by one DUE at most. Since the interference at the BS is more manageable, we consider a cellular uplink transmission system. $h_{i,B}$, $h_{i,j}$, $h_{j,B}$, $h_{j,j}$ denote the channel gain between i-th CUE and the BS, between i-th CUE and j-th DUE, between j-th DUE and the BS, between j-th DUE_T and j-th DUE_R. The channel gain is modeled as follows

$$h_{X,Y} = g_{X,Y} L_{X,Y} = g_{X,Y} \beta_{X,Y} d_{X,Y}^{-\alpha} \tag{1}$$

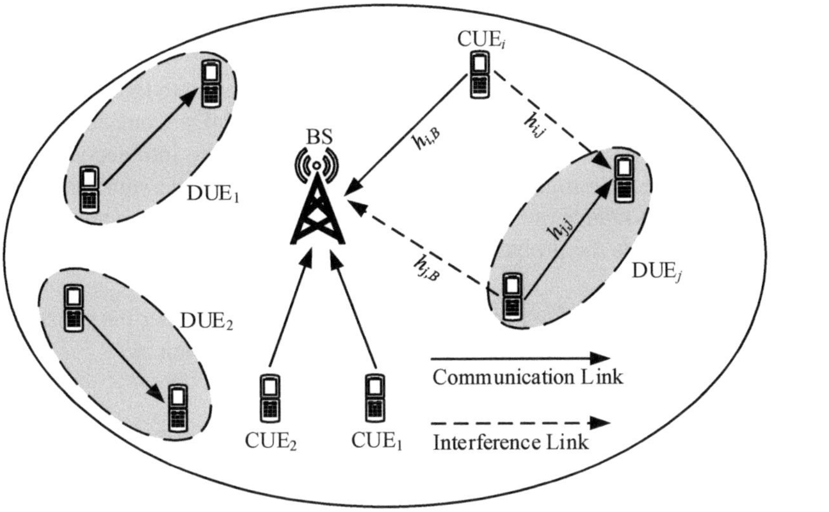

Fig. 1 System model

$g_{X,Y}$ is small-scale fading, which is assumed to be exponential distribution. $L_{X,Y}$ is the large-scale fading, $L_{X,Y} = \beta_{X,Y} d_{X,Y}^{-\alpha}$, which $d_{X,Y}^{-\alpha}$ is the path loss between X and Y, d is the distance between X and Y, α is the path loss exponent, $\beta_{X,Y}$ is shadow fading gain with log-normal distribution, i.e., $\ln(\beta_{X,Y}) \sim N(0, \sigma_\beta^2)$, where σ_β^2 is the variance of the log-normal distribution. $h_{X,Y} = \{h_{i,B}, h_{i,j}, h_{j,B}, h_{j,j}\}$.

The signal to interference and noise ratio (SINR) at the BS for the i-th CUE can be represent as

$$\gamma_i^c = \frac{p_i^c h_{i,B}}{\sum_{j=1}^{J} \rho_{i,j} p_j^d h_{j,B} + \sigma^2} \tag{2}$$

where p_i^c is the transmit power of the i-th CUE, p_j^d is the transmit power of the j-th DUE_T. $\rho_{i,j}$ is a indicator with $\rho_{i,j} = 1$ representing the j-th DUE multiplexes the i-th channel and $\rho_{i,j} = 0$ otherwise. σ^2 is noise power between transmitter and receiver. The received SINR at j-th DUE_R for j-th DUE_T can be given by

$$\gamma_j^d = \frac{p_j^d h_{j,j}}{\sum_{i=1}^{I} \rho_{i,j} p_i^c h_{i,j} + \sigma^2} \tag{3}$$

Further, we define the ergodic rate of i-th CUE to assess the average performance of the system

$$R_i = \mathbb{E}[\log_2(1 + \gamma_i^c)] \tag{4}$$

3 Problem Modeling

We propose a scheme that joint power and channel allocation to increase the ergodic rate of CUEs while guaranteeing the performance of DUEs. Considering different links may need different quality of service (QoS), we take into account the large rate for CUEs and the high reliability for DUEs. The ergodic sum rate is used to measure the performance of CUEs. The reliability of DUEs is guaranteed by outage probability, which is the probability that the received SINR γ_j^d is below a threshold γ_0^d.

In general, we maximize the ergodic sum rate of CUEs under the constraint that ensuring the reliability of DUEs. The optimal problem is given as

$$
\begin{aligned}
&\mathrm{P1}: \max_{\{p_i^c, p_j^d, \rho_{i,j}\}} \sum_{i=1}^{I} \mathbb{E}\left[\log_2\left(1+\gamma_i^c\right)\right] \\
&\text{s.t C1}: \Pr\left\{\gamma_j^d \leq \gamma_0^d\right\} \leq p_0, \quad \forall j \in J \\
&\mathrm{C2}: 0 \leq p_i^c \leq p_{\max}^c, \quad \forall i \in I \\
&\mathrm{C3}: 0 \leq p_j^d \leq p_{\max}^d, \quad \forall j \in J \\
&\mathrm{C4}: \sum_{i=1}^{I} \rho_{i,j} \leq 1, \rho_{i,j} \in \{0,1\}, \quad \forall i \in I \\
&\mathrm{C5}: \sum_{j=1}^{J} \rho_{i,j} \leq 1, \rho_{i,j} \in \{0,1\}, \quad \forall j \in J
\end{aligned} \tag{5}
$$

where P1 is the optimization function to maximize the ergodic sum rate of the CUEs. C1 is the reliability requirements for j-th DUE, where γ_0^d is the minimum SINR for reliable communications and p_0 is the maximum acceptable outage probability. C1 and C2 are the transmit power constraint, where $p_{\max}^c$ is maximal transmit power of the i-th CUE and $p_{\max}^d$ is maximal transmit power of the j-th DUE. C4 shows that the channel can only be accessed by one DUE at most, and C5 indicates that each DUE can only reuse one channel at most.

The optimization problem contains continuous variables and integer variables. It is a non-convex problem, which is usually mathematically intractable [7]. To find the optimal resource allocation, the optimization problem is divided into two steps: power allocation and channel allocation. For the constraint conditions C1, C2, C3, we find the feasible optimal transmission power for each DUE and CUE to maximize the ergodic rate under the constraint of reliability. If the power allocation is feasible, it means that the j-th DUE can reuse the i-th channel, and we treat them as candidates for channel allocation. For constraint conditions C4 and C5, we need to consider the optimal channel allocation between CUEs and DUEs. The problem of channel allocation can turn to be a bipartite matching problem, and we can use the Hungarian method to solve this matching problem.

3.1 Power Allocation

To solve the optimization problem of power distribution for arbitrary DUE and CUE, we consider an random channel reusing e.g., the j-th DUE reuses the i-th channel, $\rho_{i,j} = 1$, the optimization problem P1 is reduced as follow:

$$\begin{aligned}
&\text{P2}: \max_{\{p_i^c, p_i^d\}} \mathbb{E}\left[\log_2\left(1+\gamma_i^c\right)\right] \\
&\text{s.t C1}: \Pr\left\{\gamma_j^d \le \gamma_0^d\right\} \le p_0, \quad \forall j \in J \\
&\text{C2}: 0 \le p_i^c \le p_{\max}^c, \quad \forall i \in I \\
&\text{C3}: 0 \le p_j^d \le p_{\max}^d, \quad \forall j \in J
\end{aligned} \tag{6}$$

Substitute formula (3) into constraint C1

$$\begin{aligned}
&\Pr\left\{\gamma_j^d = \frac{p_j^d g_{j,j} L_{j,j}}{p_i^c g_{i,j} L_{i,j} + \sigma^2} \le \gamma_0^d\right\} \le p_0 \\
&= \Pr\left\{g_{j,j} \le \frac{\gamma_0^d\left(p_i^c g_{i,j} L_{i,j} + \sigma^2\right)}{p_j^d L_{j,j}}\right\} \le p_0 \\
&= \int_0^\infty \mathrm{d}g_{i,j} \int_0^{\frac{\gamma_0^d\left(p_i^c g_{i,j} L_{i,j}+\sigma^2\right)}{p_j^d L_{j,j}}} e^{-\left(g_{i,j}+g_{j,j}\right)} \mathrm{d}g_{i,j} \le p_0 \\
&= 1 - \frac{p_j^d L_{j,j} e^{-\frac{\gamma_0^d \sigma^2}{p_j^d L_{j,j}}}}{p_j^d L_{j,j} + \gamma_0^d p_i^c L_{i,j}} \le p_0
\end{aligned} \tag{7}$$

Rearranging the terms from the above inequality, p_i^c is given by

$$p_i^c \le \frac{p_j^d L_{j,j}}{\gamma_0^d L_{i,j}} \left(\frac{\exp\left(-\frac{\gamma_0^d \sigma^2}{p_j^d L_{j,j}}\right)}{1 - p_0} - 1\right) \triangleq f\left(p_j^d\right) \tag{8}$$

We can easily get the conclusion from (8) that $f(p_j^d)$ is increasing with the increasing of DUE transmit power p_j^d. Considering $p_i^c \ge 0$, so $f(p_j^d) \ge 0$, we obtain the minimal of p_j^d

$$p_j^d = \frac{-\gamma_0^d \sigma^2}{L_{j,j} \ln(1 - p_0)} \triangleq p_j^{\min} \tag{9}$$

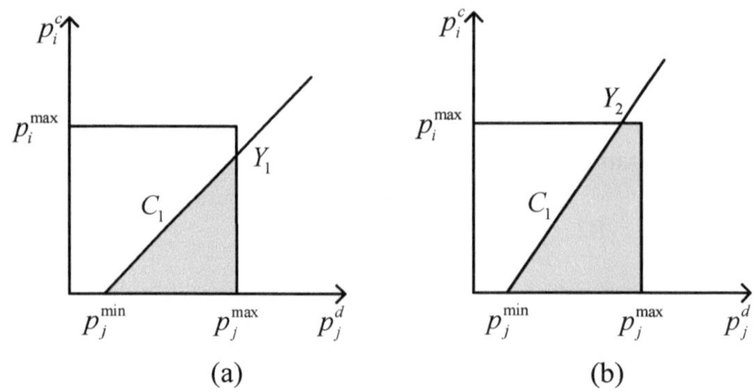

Fig. 2 Feasible region

According to the constraints C1, C2, C3, the feasible region of the power allocation is shown in the Fig. 2.

The feasible regions of the power allocation are sorted two situations due to different channel gain, maximal transmit power, predefined γ_0^d, p_0. Point $Y_1(p_j^d, f(p_j^d))$ is the intersection of straight line $p_j^d = p_j^{\max}$ and function $p_i^c = f(p_j^d)$, and point $Y_2(f^{-1}(p_i^{\max}), p_i^{\max})$ is the intersection of straight line $p_i^c = p_i^{\max}$ and function $p_i^c = f(p_j^d)$. $f^{-1}(p_j^d)$ is the inverse functions of $f(p_j^d)$, and it's closed-form expression is difficult to be obtained. Noting that $f(p_j^d)$ is a monotonic function, we can find the approximation of $f^{-1}(p_i^{\max})$ through bisection search over $f(p_j^d)$.

When the j-th DUE uses the i-th channel, the ergodic rate, $R_{i,j}(p_i^c, p_j^d)$, of i-th CUE can be written as

$$\begin{aligned}
R_{i,j}(p_i^c, p_j^d) &= \mathbb{E}[\log 2(1+\gamma_i^c)] \\
&= \mathbb{E}\left[\log_2\left(1+\frac{p_i^c g_{i,B} L_{i,B}}{p_j^d g_{j,B} L_{j,B}+\sigma^2}\right)\right] \\
&= \int_0^{\infty}\int_0^{\infty} \log_2\left(1+\frac{p_i^c g_{i,B} L_{i,B}}{p_j^d g_{j,B} L_{j,B}+\sigma^2}\right) e^{-(g_{i,B}+g_{j,B})} \mathrm{d}g_{i,B}\mathrm{d}g_{j,B}
\end{aligned} \tag{10}$$

We can draw following conclusions from (10): When p_j^d is fixed, $R_{i,j}(p_i^c, p_j^d)$ increases with the increasing of p_i^c; When p_i^c is fixed, $R_{i,j}(p_i^c, p_j^d)$ decreases with the increasing of p_j^d. Therefore, the optimal power allocation is obtained at the upper boundary of feasible region, i.e., $p_i^c = f(p_j^d)$. By substituting $p_i^c = f(p_j^d)$ into (10), γ_i^c is given by

$$\gamma_i^c = \frac{p_i^c g_{i,B} L_{i,B}}{p_j^d g_{j,B} L_{j,B}+\sigma^2}$$

$$= \frac{g_{i,B} L_{i,B} L_{j,B}}{\gamma_0^d L_{i,j}\left(g_{j,B} L_{j,B} + \frac{\sigma^2}{p_j^d}\right)} \left(\frac{\exp\left(-\frac{\gamma_0^d \sigma^2}{p_j^d L_{j,B}}\right)}{1 - p_0} - 1 \right) \tag{11}$$

It can be shown from (11) that γ_i^c increases monotonically with p_j^d in the feasible region. According to the above analysis, the scheme of optimal power distribution (p_i^{c*}, p_j^{d*}) is at the point $Y_1(p_j^d, f(p_j^d))$ for case (a) or the point $Y_2(f^{-1}(p_i^{\max}), p_i^{\max})$ for case (b) in Fig. 2.

3.2 Channel Allocation

After completing the optimal power distribution, the next step is to analyze the optimal channel allocation. When the j-th DUE multiplexes the i-th channel, the close-formed expression of the ergodic rate, $R_{i,j}(p_i^c, p_j^d)$, is given by the following equation, which is proved in Appendix.

$$\begin{aligned} R_{i,j}(p_i^c, p_j^d) &= \mathbb{E}\left[\log_2\left(1 + \gamma_i^c\right)\right] \\ &= \frac{a}{(a-b)\ln 2}\left[e^{\frac{1}{a}} E_1\left(\frac{1}{a}\right) - e^{\frac{1}{b}} E_1\left(\frac{1}{b}\right)\right] \end{aligned} \tag{12}$$

where $a = \frac{p_i^c L_{i,B}}{\sigma^2}$, $b = \frac{p_j^d L_{j,B}}{\sigma^2}$ and $E_1(x) = \int\limits_0^x \frac{e^{-t}}{t} \mathrm{d}t$.

The ergodic rate under the optimal power allocation is $R_{i,j}^* = R_{i,j}(p_i^{c*}, p_j^{d*})$, and the optimization problem P1 is transformed to P3

$$\begin{aligned} \text{P3} : &\max_{\{\rho_{i,j}\}} \sum_{i=1}^{I} \sum_{j=1}^{J} \rho_{i,j} R_{i,j}^* \\ \text{s.t C4} : &\sum_{i=1}^{I} \rho_{i,j} \le 1, \rho_{i,j} \in \{0, 1\} \\ \text{C5} : &\sum_{i=1}^{J} \rho_{i,j} \le 1, \rho_{i,j} \in \{0, 1\} \end{aligned} \tag{13}$$

According to the previous assumptions, each DUE can only access one channel and each channel can only be reused by one DUE at most. The optimization problem turns to be a maximum weight bipartite matching problem, and we can resolve it by using the Hungarian method [8].

3.3 Time Complexity

From the above analysis, we find the optimal scheme of joint power allocation and channel allocation for the proposed system model. Supposing that we have an accuracy of ζ for the optimal power allocation $f^{-1}(p_i^{\max})$, the bisection search need $\log(1/\zeta)$ iterations. The total time complexity for computing all CUEs and DUEs is given by $O(I \times J \times \log(1/\zeta))$. The optimal channel allocation is obtained by using the Hungarian method in $O(I^3)$ time. Finally, the total time complexity of the proposed scheme of resource allocation is $O(I \times J \times \log(1/\zeta) + I^3)$.

4 Simulation Results

Numerical results are presented in this section to analyze the scheme of resource allocation for this system. The CUEs and DUEs are located around of BS by spatial poison process, and the DUE_T chooses the closet receiver as it's DUE_R. The major simulation parameters are list in Table 1 by default.

Figure 3 shows the effect of varying the reliability of DUEs on the ergodic sum rate for $p_{\max}^d = p_{\max}^c = 20$ dBm and $p_{\max}^d = p_{\max}^c = 24$ dBm. It can be easily shown that the ergodic sum rate of CUEs will increase when the higher outage probability of DUEs is allowed. The high outage probability means that DUEs can tolerance higher interference, thus urging CUEs to increase their transmit power. Therefore, the ergodic sum rate of CUEs will increase. It also illustrates that the high maximum transmit power will increase the ergodic sum rate.

Table 1 Simulation parameters

Parameter	Value
CUEs location	Spatial Poisson process
DUEs location	Spatial Poisson process
Cell radius	500 m
Fast fading g	$E(1)$
Shadowing fading β	$\sigma_\beta^2 = 8$ dBm
Path loss exponent α	3
Maximum CUE transmit power $p_{\max}^c$	20 dBm, 24 dBm
Maximum DUE transmit power $p_{\max}^d$	20 dBm, 24 dBm
Number of DUEs J	20
Number of CUEs I	20
Noise power σ^2	-114 dBm
SINR threshold for DUE γ_0^d	5 dB
Reliability of DUE p_0	10^{-3}

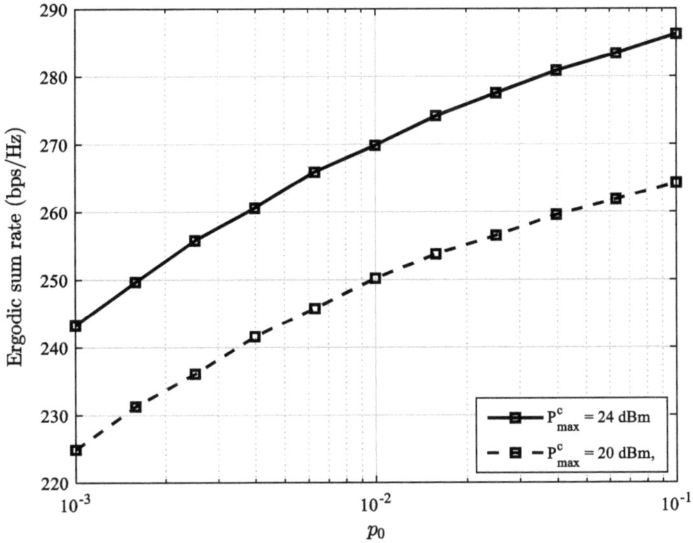

Fig. 3 The ergodic sum rate versus the reliability of DUEs

Figure 4 depicts the influence between the ergodic sum rate and the different number of DUEs. J/I is the ratio of the number of DUEs and CUEs when $I = 20$. It can be seen that increasing the number of DUEs leads to reducing the preference of the ergodic sum rate of CUEs, because more DUEs will cause more interference

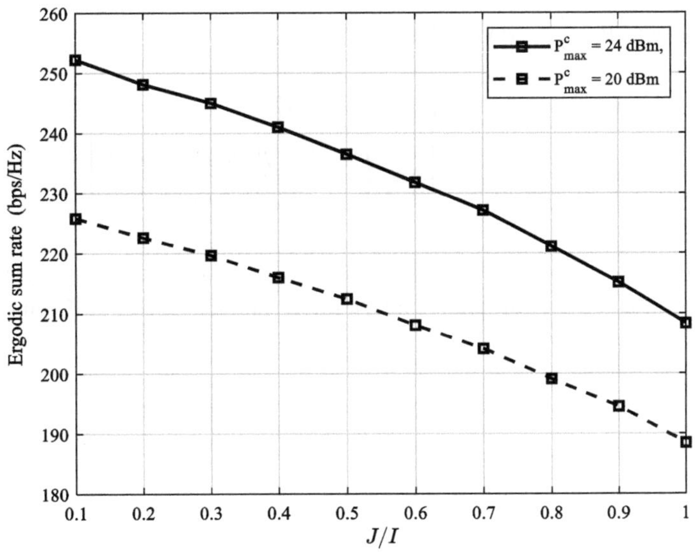

Fig. 4 The ergodic sum rate versus the number of DUEs

to CUEs, so the SINR of CUEs will decrease. It also illustrates that increasing the maximum transmit power will increase the ergodic sum rate of CUEs.

5 Conclusion

The scheme of joint power allocation and channel distribution has been raised to improve the preference of CUEs under constraint of the reliability of DUEs. Assuming that the large rate demand for CUEs and the acceptable outage probability for DUEs, we maximize the ergodic sum rate of CUEs while assuring the outage preference of DUEs. The optimization problem is hardly solved mathematically, so we transfer this optimal problem into two steps: power allocation and channel allocation. Simulation results demonstrate that this scheme of resource allocation can increase the performance of CUEs under acceptable interference to the original network.

In this paper, we suppose that each DUE can only access one channel, each channel can only be reused by one DUE at most. In the future work, we can research that one DUE access multiple channels or one channel can be reused by different DUEs. It is also interesting to research that multiple channels can be accessed by multiple DUEs.

Appendix

The ergodic rate $R_{i,j}(p_i^c, p_j^d)$ can be calculated as

$$\begin{aligned} R_{i,j}(p_i^c, p_j^d) &= \mathbb{E}\left[\log_2\left(1+\gamma_i^c\right)\right] \\ &= \mathbb{E}\left[\log_2\left(1+\frac{p_i^c g_{i,B} L_{i,B}}{p_j^d g_{j,B} L_{j,B}+\sigma^2}\right)\right] \end{aligned} \tag{14}$$

We define $a = \frac{p_i^c L_{i,B}}{\sigma^2}$, $b = \frac{p_j^d L_{j,B}}{\sigma^2}$, $X = g_{i,B}$, $Y = g_{j,B}$, $Z = \frac{aX}{bY+1}$. Assuming $g_{i,B}$ $E(1)$, $g_{j,B}$ $E(1)$, the cumulative distribution function (CDF) of Z can be written as

$$\begin{aligned} F_Z(z) &= \Pr\left\{\frac{a g_{i,B}}{b g_{j,B}+1} \le z\right\} \\ &= \int_0^{\infty} \mathrm{d}y \int_0^{\frac{z(1+by)}{a}} e^{-(x+y)} \mathrm{d}x \end{aligned} \tag{15}$$

Then, the close-formed expression of the ergodic rate can be given by

$$
\begin{aligned}
R_{i,j}\left(p_i^c, p_j^d\right) &= E\left[\log_2(1+Z)\right] \\
&= \frac{1}{\ln 2}\int_0^\infty \ln(1+z) f_Z(z)\mathrm{d}z \\
&= \frac{a}{(a-b)\ln 2}\left[\int_0^\infty \frac{e^{-\frac{z}{a}}}{z+1}\mathrm{d}z - \int_0^\infty \frac{e^{-\frac{z}{a}}}{z+\frac{a}{b}}\mathrm{d}z\right] \\
&= \frac{a}{(a-b)\ln 2}\left[e^{\frac{1}{a}} E_1\left(\frac{1}{a}\right) - e^{\frac{1}{b}} E_1\left(\frac{1}{b}\right)\right]
\end{aligned} \tag{16}
$$

References

1. Kuang, Z., Liu, G., Li, G., et al.: Energy efficient resource allocation algorithm in energy harvesting-based D2D heterogeneous networks. IEEE Internet Things J. **6**(1), 557–567 (2019)
2. Lee, J., Lee, J.H.: Performance analysis and resource allocation for cooperative D2D communication in cellular networks with multiple D2D pairs. IEEE Commun. Lett. **23**(5), 909–912 (2019)
3. Sun, Z., Yang, D.: A D2D wireless resource allocation scheme based on overall fairness. 3D Res. **10**(2) (2019)
4. Lee, N., Lin, X., Andrews, J.G., et al.: Power control for D2D underlaid cellular networks: modeling, algorithms, and analysis. IEEE J. Sel. Areas Commun. **33**(1), 1–13 (2015)
5. Chen, C.-Y., Sung, C.-A., Chen, H.-H.: Capacity maximization based on optimal mode selection in multi-mode and multi-pair D2D communications. IEEE Trans. Veh. Technol. **68**(7), 6524–6534 (2019)
6. Radaydeh, R.M., Al-Qahtani, F.S., Celik, A., et al.: Generalized imperfect D2D associations in spectrum-shared cellular networks under transmit power and interference constraints. IEEE Access **8**, 182517–182536 (2020)
7. Liang, L., Li, G.Y., Xu, W.: Resource allocation for D2D-enabled vehicular communications. IEEE Trans. Commun. **65**(7), 3186–3197 (2017)
8. Lim, D.-W., Kang, J., Chun, C.-J., et al.: Joint transmit power and time-switching control for device-to-device communications in SWIPT cellular networks. IEEE Commun. Lett. **23**(2), 322–325 (2019)

$l_{1/2}$-SVD Based Channel Estimation for MmWave Massive MIMO

Xiaoli Jing, Xianpeng Wang, Xiang Lan, and Liangtian Wan

1 Introduction

mmWave massive MIMO technology has the advantages of ultra-high transmission rate, large transmission bandwidth and lower transmission delay, it has become one of the important development trends of next generation mobile communication [1]. However, mmWave massive MIMO still faces technical problems such as serious reflection loss, multipath delay, and easy blocking interruption. These problems have brought challenges to channel estimation [2].

In recent years, the channel estimation algorithms based on compressed sensing (CS) mainly include greedy algorithms and convex optimization algorithms. Orthogonal Matching Pursuit (OMP) is a representative algorithm in greedy algorithms [3, 4]. But the greedy estimation algorithm is more likely to fall into the local optimal solution. Another type of recovery algorithm is to construct the sparse recovery problem as a l_0-norm optimization problem. It is more difficult to find the optimal result. Therefore, the convex optimization estimation algorithm is usually used to approximate. [5] proposed a l_1-norm-based channel estimation scheme, which reconstructs the problem of CS. However, in practice, due to the influence of random noise, the sparsest solution cannot be obtained in the l_1-norm solution. Rong et al. [6] has clearly pointed out that $l_q(0 < q < 1)$-norm has obtained a more sparse solution, but the quantization of the angle may introduce errors, so the channel estimation algorithm needs further improvement. In [7], an objective function based on $l_{1/2}$-regularization was constructed, and then the super-resolution channel estimation

X. Jing · X. Wang (✉) · X. Lan
State Key Laboratory of Marine Resource Utilization in South China Sea and School of Information and Communication Engineering, Hainan University, Haikou 570228, China
e-mail: wxpeng1986@126.com

L. Wan
School of Software, Dalian University of Technology, Dalian 116024, China

L. C. Jain et al. (eds.), *Advances in Wireless Communications and Applications*, Smart Innovation, Systems and Technologies 299,
https://doi.org/10.1007/978-981-19-2255-8_2

was finally realized through iterative optimization. However, the method proposed in this document is still unable to achieve the desired effect in terms of complexity.

The paper proposes a novel mmWave massive MIMO channel estimation algorithm. First, an objective function based on $l_{1/2}$-regularization is constructed. Then, the channel estimation problem is transformed into an alternative optimization problem through the gradient descent method, and the optimal angle parameter estimation value is obtained.

2 System Model

Under the system of hybrid-precoding mmWave massive MIMO, the transmit antennas is equipped with N_t antennas, the receiving end is equipped with N_r antennas. The number of transmitter RF chains and receiver RF chains are N_t^{RF} and N_r^{RF}, respectively. And both the transmitting end and the receiving end are single-stream communication. The received signal is:

$$\boldsymbol{Y} = \boldsymbol{W}^{H}\boldsymbol{H}\boldsymbol{P}\boldsymbol{s} + \boldsymbol{n} \tag{1}$$

where $\boldsymbol{Y}$ is the receiving signal of the system, $\boldsymbol{W}$ is the hybrid combination matrix at the receiving end, $\boldsymbol{P}$ is the hybrid precoding matrix, $\boldsymbol{H}$ is the channel matrix, $\boldsymbol{s}$ is the pilot signal at the transmitter, $\boldsymbol{n}$ is the combined received Gaussian white noise.

The paper adopts the widely used Saleh-Valenzuela channel model

$$\boldsymbol{H} = \sqrt{\frac{N_t N_r}{L}} \sum_{l=1}^{L} \beta_l \mathbf{a}(\theta_{r,l})\mathbf{a}^H(\theta_{t,l}) \tag{2}$$

where L is the effective propagation path ($L \ll \min(N_r, N_\mathrm{t})$), β_l is the complex gain of the l-th path, $\theta_{r,l}$ and $\theta_{t,l}$ are the corresponding arrival angle and transmit angle, respectively. $\mathbf{a}(\theta_{r,l})$ and $\mathbf{a}(\theta_{t,l})$ can be expressed as

$$\mathbf{a}(\theta_r) = \frac{1}{\sqrt{N_t}}\left[1, e^{j2\pi d \sin\theta_t/\lambda}, \ldots, e^{j2\pi(N-1)d \sin\theta_t/\lambda}\right]^T \tag{3}$$

$$\mathbf{a}(\theta_t) = \frac{1}{\sqrt{N_t}}\left[1, e^{j2\pi d \sin\theta_t/\lambda}, \ldots, e^{j2\pi(N-1)d \sin\theta_t/\lambda}\right]^T \tag{4}$$

where d = $\frac{\lambda}{2}$ is the distance between adjacent elements.

Therefore, the mmWave channel $\boldsymbol{H}$ can also be expressed as

$$\boldsymbol{H} = \boldsymbol{A}(\boldsymbol{\theta}_r)\boldsymbol{\beta}\boldsymbol{A}^{\mathrm{H}}(\boldsymbol{\theta}_t) \tag{5}$$

where $\boldsymbol{\beta} = \sqrt{\frac{N_t N_r}{L}} diag[\beta_1, \cdots, \beta_L]$.

Using $\mathbf{x}= \boldsymbol{Ps}$ to represent a pilot signal transmitted, the i-th element in vector $\boldsymbol{x}$ corresponds to the signal sent by the i-th transmitting antenna. The precoding matrix and the transmitted signal content $\mathrm{tr}(\boldsymbol{P}\boldsymbol{P}^{\mathrm{H}}) \le \rho$ and $\mathrm{E}(\boldsymbol{s}\boldsymbol{s}^{\mathrm{H}}) = \boldsymbol{I}_N$, respectively. The received pilot signal can also be expressed as

$$\boldsymbol{Y} = \boldsymbol{U}^{\mathrm{H}}\boldsymbol{H}\boldsymbol{X} + \boldsymbol{N} \tag{6}$$

Due to the sparse nature of the channel, the sparse channel estimation problem can be transformed into

$$\min_{\hat{\boldsymbol{\beta}},\hat{\boldsymbol{\theta}}_r,\hat{\boldsymbol{\theta}}_t} \left\| \hat{\boldsymbol{\beta}} \right\|_0, \text{ s.t.} \left\| \boldsymbol{Y} - \boldsymbol{U}^{\mathrm{H}}\hat{\boldsymbol{H}}\boldsymbol{X} \right\|_{\mathrm{F}} \le \varepsilon \tag{7}$$

ε is a threshold set to control the estimation error.

3 Description of the Proposed Channel Estimation Algorithm

3.1 Proposed Optimization Alternative Formula

Generally speaking, the optimization of l_0-norm is difficult to solve, so in most researches, l_1-norm is often replaced by l_0-norm. However, in practice, due to the influence of random noise, a non-sparse solution is formed in the process of solving the l_1-norm. Therefore, this paper chooses a new regular term $l_{1/2}$-norm with stronger anti-noise ability to obtain a more sparse solution. The reason for choosing the $l_{1/2}$-norm is that the regular term we need is easier to solve than the l_0-norm, and at the same time obtain a sparser solution than the l_1-norm [7]. The sparse representation ability of the $l_q(0 < q < 1/2)$-norm is equivalent to that of the $l_{1/2}$-norm, and the $l_q(1/2 < q < 1)$-norm is weaker than the $l_{1/2}$-norm. Replacing the l_0-norm in the above formula with $l_{1/2}$-norm to get

$$\min_{\boldsymbol{\beta},\boldsymbol{\theta}_r,\boldsymbol{\theta}_t} F(\boldsymbol{\beta}) = \|\boldsymbol{\beta}\|_{1/2}, \text{ s.t.} \left\| \boldsymbol{Y} - \boldsymbol{U}^{\mathrm{H}}\hat{\boldsymbol{H}}\boldsymbol{X} \right\|_{\mathrm{F}} \le \varepsilon \tag{8}$$

γ is introduced to control the error between sparsity and data fitting. The problem (8) can be refactored into the following form

$$\min_{\boldsymbol{\beta},\boldsymbol{\theta}_r,\boldsymbol{\theta}_t} G(\boldsymbol{\beta}, \boldsymbol{\theta}_r, \boldsymbol{\theta}_t) = \|\boldsymbol{\beta}\|_{1/2} + \gamma \left\| \boldsymbol{Y} - \boldsymbol{U}^{\mathrm{H}}\hat{\boldsymbol{H}}\boldsymbol{X} \right\|_F^2 \tag{9}$$

The $l_q(0 < q < 1)$-norm has non-convex characteristic, and its solution can be transformed into an iterative convex optimization process, which is a form of equivalent replacement. The specific expression is

$$\min_{\boldsymbol{\beta},\boldsymbol{\theta}_r,\boldsymbol{\theta}_t} S^{(t)}(\boldsymbol{\beta},\boldsymbol{\theta}_r,\boldsymbol{\theta}_t) \triangleq \sum_{i=1}^{L}\left(\left(\boldsymbol{\beta}_i^{(t)}\right)^2+\delta\right)^{-3/4}\boldsymbol{\beta}_i^2+\gamma\left\|\boldsymbol{Y}-\boldsymbol{U}^{\mathrm{H}}\hat{\boldsymbol{H}}\boldsymbol{X}\right\|_F^2 \tag{10}$$

Based on the above statements, this paper constructs an iterative proxy function for formula (10). Then the solution $G(\boldsymbol{\beta},\boldsymbol{\theta}_r,\boldsymbol{\theta}_t)$ can be converted into an optimization problem of substitution function [6].

$$\min_{\boldsymbol{\beta},\boldsymbol{\theta}_r,\boldsymbol{\theta}_t} S^{(t)}(\boldsymbol{\beta},\boldsymbol{\theta}_r,\boldsymbol{\theta}_t) \triangleq \boldsymbol{\beta}^H\boldsymbol{D}^{(t)}\boldsymbol{\beta}+\gamma\left\|\boldsymbol{Y}-\boldsymbol{U}^{\mathrm{H}}\hat{\boldsymbol{H}}\boldsymbol{X}\right\|_F^2 \tag{11}$$

where

$$\boldsymbol{D}^{(t)} \triangleq diag\left[\frac{1}{\left(\left(\hat{\boldsymbol{\beta}}_{i1}^{(t)}\right)^2+\delta\right)^{3/4}}\frac{1}{\left(\left(\hat{\boldsymbol{\beta}}_{i2}^{(t)}\right)^2+\delta\right)^{3/4}}\cdots\frac{1}{\left(\left(\hat{\boldsymbol{\beta}}_{iL}^{(t)}\right)^2+\delta\right)^{3/4}}\right] \tag{12}$$

In (10), we will encounter a situation. This situation is when $\boldsymbol{\beta}_i^{(t)}=0$, if δ is not introduced, (10) will be undefined. Therefore, in the alternative optimization process, this article not only needs to introduce δ. In order to obtain better estimation performance, the parameters will gradually decrease in the iterative process instead of a fixed value [8].

In the t-th iteration, $\hat{\boldsymbol{\beta}}^{(t+1)}$, $\hat{\boldsymbol{\theta}}_r^{(t+1)}$, and $\hat{\boldsymbol{\theta}}_t^{(t+1)}$ will be found so that $S^{(t)}(\boldsymbol{\beta},\boldsymbol{\theta}_r,\boldsymbol{\theta}_t)$ satisfies the following inequality

$$S^{(t)}(\hat{\boldsymbol{\beta}}^{(t+1)},\hat{\boldsymbol{\theta}}_r^{(t+1)},\hat{\boldsymbol{\theta}}_t^{(t+1)}) \le S^{(t)}(\hat{\boldsymbol{\beta}}^{(t)},\hat{\boldsymbol{\theta}}_r^{(t)},\hat{\boldsymbol{\theta}}_t^{(t)}) \tag{13}$$

Combining (9), (10) and (11), we have

$$\begin{aligned} &G\left(\hat{\boldsymbol{\beta}}^{(t+1)},\hat{\boldsymbol{\theta}}_r^{(t+1)},\hat{\boldsymbol{\theta}}_t^{(t+1)}\right)-S^{(t)}\left(\hat{\boldsymbol{\beta}}^{(t+1)},\hat{\boldsymbol{\theta}}_r^{(t+1)},\hat{\boldsymbol{\theta}}_t^{(t+1)}\right)\\ &=F\left(\hat{\boldsymbol{\beta}}^{(t+1)}\right)-\sum_{l=1}^{L}\frac{\left|\hat{\boldsymbol{\beta}}^{(t+1)}\right|^2}{\left(\left|\hat{\boldsymbol{\beta}}^{(t+1)}\right|^2+\delta\right)^{3/4}}\end{aligned}$$

$$\le F\left(\hat{\boldsymbol{\beta}}^{(t)}\right) - \sum_{l=1}^{L} \frac{\left|\hat{\boldsymbol{\beta}}^{(t)}\right|^2}{\left(\left|\hat{\boldsymbol{\beta}}^{(t)}\right|^2 + \delta\right)^{3/4}}$$

$$= G\left(\hat{\boldsymbol{\beta}}^{(t)}, \hat{\boldsymbol{\theta}}_r^{(t)}, \hat{\boldsymbol{\theta}}_t^{(t)}\right) - S^{(t)}\left(\hat{\boldsymbol{\beta}}^{(t)}, \hat{\boldsymbol{\theta}}_r^{(t)}, \hat{\boldsymbol{\theta}}_t^{(t)}\right) \tag{14}$$

It is worth noting that when $\boldsymbol{\beta} = \hat{\boldsymbol{\beta}}^{(t)}$, $G(\boldsymbol{\beta}, \boldsymbol{\theta}_r, \boldsymbol{\theta}_t) - S^{(}\mathrm{t})(\boldsymbol{\beta}, \boldsymbol{\theta}_r, \boldsymbol{\theta}_t)$ gets the maximum. And we can get

$$G(\hat{\boldsymbol{\beta}}^{(t+1)}, \hat{\boldsymbol{\theta}}_r^{(t+1)}, \hat{\boldsymbol{\theta}}_t^{(t+1)})$$
$$= G(\hat{\boldsymbol{\beta}}^{(t+1)}, \hat{\boldsymbol{\theta}}_r^{(t+1)}, \hat{\boldsymbol{\theta}}_t^{(t+1)}) - S^{(t)}(\hat{\boldsymbol{\beta}}^{(t+1)}, \hat{\boldsymbol{\theta}}_r^{(t+1)}, \hat{\boldsymbol{\theta}}_t^{(t+1)})$$
$$+ S^{(t)}(\hat{\boldsymbol{\beta}}^{(t+1)}, \hat{\boldsymbol{\theta}}_r^{(t+1)}, \hat{\boldsymbol{\theta}}_t^{(t+1)})$$
$$\le G(\hat{\boldsymbol{\beta}}^{(t)}, \hat{\boldsymbol{\theta}}_r^{(t)}, \hat{\boldsymbol{\theta}}_t^{(t)}) - \left[S^{(t)}(\hat{\boldsymbol{\beta}}^{(t)}, \hat{\boldsymbol{\theta}}_r^{(t)}, \hat{\boldsymbol{\theta}}_t^{(t)}) - S^{(t)}(\hat{\boldsymbol{\beta}}^{(t+1)}, \hat{\boldsymbol{\theta}}_r^{(t+1)}, \hat{\boldsymbol{\theta}}_t^{(t+1)})\right]$$
$$\le G(\hat{\boldsymbol{\beta}}^{(t)}, \hat{\boldsymbol{\theta}}_r^{(t)}, \hat{\boldsymbol{\theta}}_t^{(t)}) \tag{15}$$

To simplify $S^{(t)}(\boldsymbol{\beta}, \boldsymbol{\theta}_r, \boldsymbol{\theta}_t)$, this algorithm constructs two functions. One is to use $\boldsymbol{\theta}_r$ and $\boldsymbol{\theta}_t$ to represent the correlation function of $\boldsymbol{\beta}$, and the other is to use $\boldsymbol{\theta}_r$ and $\boldsymbol{\theta}_t$ to represent the function of S. The specific expression is as follows

$$\boldsymbol{\beta}_{opt}^{(t)} \triangleq arg\min_{\boldsymbol{\beta}} S^{(t)}(\boldsymbol{\beta}, \boldsymbol{\theta}_r, \boldsymbol{\theta}_t) = \left(\gamma^{-1}\boldsymbol{D}^{(t)} + \boldsymbol{K}^H\boldsymbol{K}\right)^{-1}\left(\boldsymbol{K}^H\boldsymbol{Y}\right) \tag{16}$$

where

$$K = \boldsymbol{U}^H\boldsymbol{A}(\boldsymbol{\theta}_r)diag\left(\boldsymbol{A}^H(\boldsymbol{\theta}_t)\boldsymbol{X}\right) \tag{17}$$

Finally, substituting (16) into (11), it converts $S^{(t)}(\boldsymbol{\beta}, \boldsymbol{\theta}_r, \boldsymbol{\theta}_t)$ into a function only related to the angle parameter.

$$S_{opt}^{(t)}(\boldsymbol{\theta}_r, \boldsymbol{\theta}_t) \triangleq \min_{\boldsymbol{\beta}} S^{(t)}(\boldsymbol{\beta}, \boldsymbol{\theta}_r, \boldsymbol{\theta}_t) = -\left(\boldsymbol{K}^H\boldsymbol{Y}\right)^H\left(\gamma^{-1}\boldsymbol{D}^{(t)}\boldsymbol{K}^H\boldsymbol{K}\right)^{-1}$$
$$+ \left(\boldsymbol{K}^H\boldsymbol{Y}\right) + \boldsymbol{Y}^H\boldsymbol{Y} \tag{18}$$

After that, in (5), we only need to estimate $\boldsymbol{\theta}_r$ and $\boldsymbol{\theta}_t$.

3.2 Channel Estimation Based on IR

In order to solve the problem of angle parameter estimation the IR-based channel estimation method is used in this paper.

γ is used to adjust the weight between $\boldsymbol{\beta}^H \boldsymbol{D}^{(t)} \boldsymbol{\beta}$ and $\|\boldsymbol{Y} - \boldsymbol{K}\boldsymbol{\beta}\|_F^2$. γ will be updated in the following ways

$$\gamma = \min(d/r^{(t)}, \gamma_{\max}) \tag{19}$$

where $\gamma_{\max}$ is used to ensure the good operation of the algorithm. $r^{(t)}$ is the residual square of the previous iteration.

$$r^{(t)} = \left\| \boldsymbol{Y} - \boldsymbol{U}^H \boldsymbol{A}(\hat{\boldsymbol{\theta}}_r^{(t)}) \hat{\boldsymbol{\beta}}^{(t)} \boldsymbol{A}^H (\hat{\boldsymbol{\theta}}_t^{(t)}) \boldsymbol{X} \right\|_F^2 \tag{20}$$

The algorithm uses gradient descent method to estimate the angle parameters. The method is expressed as follows

$$\begin{aligned} \hat{\boldsymbol{\theta}}_r^{(t+1)} &= \hat{\boldsymbol{\theta}}_r^{(t)} - \varsigma \cdot \nabla_{\boldsymbol{\theta}_r} S_{opt}^{(t)}(\hat{\boldsymbol{\theta}}_r^{(t)}, \hat{\boldsymbol{\theta}}_t^{(t)}) \\ \hat{\boldsymbol{\theta}}_t^{(t+1)} &= \hat{\theta}_t^{(t)} - \varsigma \cdot \nabla_{\boldsymbol{\theta}_t} S_{opt}^{(t)}(\hat{\boldsymbol{\theta}}_r^{(t)}, \hat{\boldsymbol{\theta}}_t^{(t)}) \end{aligned} \tag{21}$$

where ∇ is the gradient operator, ς is the step size.

In this algorithm, a SVD-based scheme is used to initialize the angle parameters. The received signal $\boldsymbol{Y}$ is simplified by SVD. We have $\boldsymbol{Y} = \boldsymbol{W}\Sigma\boldsymbol{V}^H$, Σ is a diagonal matrix. The angle parameter initialization formula is expressed as

$$\begin{aligned} \hat{\boldsymbol{\theta}}_r^{(0)} &= \operatorname{argmin} \boldsymbol{W}^H \boldsymbol{U}^H \boldsymbol{A}(\boldsymbol{\theta}_r) \\ \hat{\boldsymbol{\theta}}_t^{(0)} &= \operatorname{argmin} \boldsymbol{V}^H \boldsymbol{X}^H \boldsymbol{A}(\boldsymbol{\theta}_t) \end{aligned} \tag{22}$$

The pre-processing method based on SVD can debase the complexity and at the same time find the angular domain grid closest to the real AoA/AoD.

Algorithm 1. The specific Flow of the Algorithm

Input	The receive signal $\boldsymbol{Y}$; Initialize the angle parameter $\hat{\boldsymbol{\theta}}_{\mathrm{r}}^{(0)}$ and $\hat{\boldsymbol{\theta}}_{\mathbf{t}}^{(0)}$; Delete threshold β_{th}; Fault tolerance threshold $\varepsilon_{\mathrm{th}}$
Step 1	Initialize $\hat{\boldsymbol{\beta}}^{(0)} = \boldsymbol{\beta}_{opt}(\hat{\boldsymbol{\theta}}_{\mathbf{r}}^{(0)}, \hat{\boldsymbol{\theta}}_{\mathbf{t}}^{(0)})$ according to (16)
Step 2	Update γ by (20)
Step 3	Calculated $\boldsymbol{S}_{\mathrm{opt}}^{(}\mathrm{t})(\boldsymbol{\theta}_{\mathbf{r}}, \boldsymbol{\theta}_{\mathbf{t}})$ by (18)
Step 4	Estimate the new $\hat{\boldsymbol{\theta}}_{\mathbf{r}}^{(\mathrm{t}+1)}$ and $\hat{\boldsymbol{\theta}}_{\mathbf{t}}^{(\mathrm{t}+1)}$ by (22)

(continued)

(continued)

Input	The receive signal $\boldsymbol{Y}$; Initialize the angle parameter $\hat{\boldsymbol{\theta}}_{\mathrm{r}}^{(0)}$ and $\hat{\boldsymbol{\theta}}_{\mathbf{t}}^{(0)}$; Delete threshold β_{th}; Fault tolerance threshold $\varepsilon_{\mathrm{th}}$
Step 5	Estimate $\hat{\boldsymbol{\beta}}^{(\mathrm{t}+1)}$ by (16); If $\hat{\boldsymbol{\beta}}_{l}^{(\mathrm{t}+1)} < \beta_{\mathrm{th}}$, then trim the path
Step 6	Until $L^{(\mathrm{t})} = L^{(\mathrm{t}+1)}$ and $\left\| \hat{\boldsymbol{\beta}}^{(\mathrm{t}+1)} - \hat{\boldsymbol{\beta}}^{(\mathrm{t})} \right\|_2 < \varepsilon_{\mathrm{th}}$, the iteration ends

4 Simulation Results

The properties of the proposed algorithm is verified through some simulation comparison results in this section.

The simulation parameter is set to $N_{\mathrm{t}} = N_{\mathrm{r}} = 64$. We use four algorithms for comparison, including LS-based channel estimation, OMP-based channel estimation, ADMM-based channel estimation [9], and Cramer-Rao bound (CRB).

Figure 1a shows the NMSE for channel estimation of various algorithms under different SNRs. The accuracy of each algorithm increases as the SNR increases. The performance of the LS-based algorithm has the worst estimation accuracy and almost no change, indicating that the traditional LS is not suitable for mmWave massive MIMO channel estimation. This is mainly due to the easy attenuation characteristics of mmWave, and its estimated environment is often low SNR. From the overall effect, the proposed algorithm is closer to CRB.

The effect of path numbers on the NMSE of the four algorithms is shown in Fig. 1b. Figure 1b shows that with the number of paths increases, the estimation accuracy of the three algorithms show a downward trend. Under the same number of

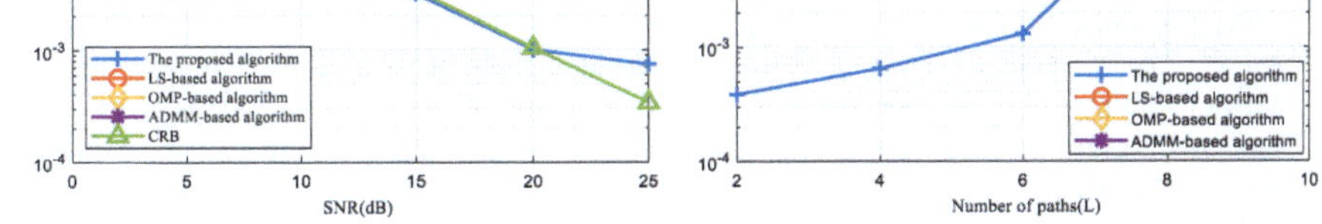

(a) (b)

Fig. 1 **a** NMSE versus SNR. **b** NMSE versus L

paths, the channel estimation performance of the proposed algorithm is better than the other algorithms.

In summary, the traditional channel estimation method is not suitable for mm-Wave massive MIMO system to a certain extent. The proposed algorithm based on CS becomes a better choice.

5 Conclusion

In short, a channel estimation method based on $l_{1/2}$-SVD is proposed. The basic idea of the algorithm is to transform the channel estimation into the recovery of sparse signals. The iterative replacement function based on $l_{1/2}$ is constructed first, and then preprocessed by SVD, which reduces the computational complexity. Finally, the objective function is optimized by the gradient descent method to obtain the optimal solution of the angle parameter. After simulation analysis, it can be gained that the proposed algorithm has certain advantages and provides guidance for subsequent channel estimation algorithm research.

Acknowledgements This work was supported by the National Natural Science Foundation of China (No. 61861015 and 61961013), Key Research and Development Program of Hainan Province (No. ZDYF2019011), National Key Research and Development Program of China (No. 2019CXTD400), Young Elite Scientists Sponsorship Program by CAST (No. 2018QNRC001), and the Scientific Research Setup Fund of Hainan University (No. KYQD(ZR) 1731).

References

1. Cong, J., Wang, X., Huang, M., Wan, L.: Robust DOA estimation method for MIMO radar via deep neural networks. IEEE Sens. J. **21**(6), 7498–7507 (2020)
2. Wang, X., Yang, L.T., Meng, D., Dong, M., Ota, K., Wang, H.: Multi-UAV cooperative localization for marine targets based on weighted subspace fitting in SAGIN environment. IEEE Internet Things J. (2021)
3. Wang, X., Wan, L., Huang, M., Shen, C., Han, Z., Zhu, T.: Low-complexity channel estimation for circular and noncircular signals in virtual MIMO vehicle communication systems. IEEE Trans. Veh. Technol. **69**, 3916–3928 (2021)
4. Shao, J., Wang, X., Lan, X., et al.: GAMP-SBL-based channel estimation for millimeter-wave MIMO systems. EURASIP J. Adv. Signal Process. **1**, 1–22 (2021)
5. Hu, C., Dai, L., Gao, Z., Fang, J.: Super-resolution channel estimation for MmWave massive MIMO with hybrid precoding. IEEE Trans. Veh. Technol. **67**, 8954–8958 (2018)
6. Zhang, Z., Liang, Y., Gui, G.: -Regularization based sparse channel estimation for MmWave massive MIMO systems. In: IEEE 23rd International Conference on Digital Signal Processing (2018)
7. Zhang, Z., Liang, Y., Shi, W., Yuan, L., Gui, G.: Regularization based super-resolution sparse channel estimation for MmWave massive MIMO systems. IEEE Access **7**, 75837–75844 (2019)
8. Wang, X., Huang, M., Wan, L.: Joint 2D-DOD and 2D-DOA estimation for coprime EMVS–MIMO radar. In: Circuits, Systems, and Signal Processing (2021)

9. Vlachos, E., Alexandropoulos, G.C., Thompson, J.: Massive MIMO channel estimation for millimeter wave systems via matrix completion. IEEE Signal Process. Lett. **25**(11), 1675–1679 (2018)

Enhancing Physical Layer Security of Cooperative NOMA Systems Using User Selection

Menghuan Ma, Yucheng He, Yan Zhang, Yu Zhang, and Lin Zhou

Abstract In this paper, we study a physical layer security improvement method for a multi-user cooperative non-orthogonal multiple access (NOMA) system in the presence of an adaptive eavesdropper. In the system, the eavesdropper adaptively switches the eavesdropping mode and interference mode. In order to resist the eavesdropping attack of the eavesdropper, a number of device-to-device (D2D) relay users are available for cooperating to forward legitimate signals, and one of them can be selected for assistance through a user selection strategy based on the channel side information (CSI) of the eavesdropping channel. The selected relay uses the NOMA technology to superimpose ordinary signals that only require data services with the received sensitive signals, and then forwards the superimposed signals. The ergodic secrecy sum capacity lower bound of the system is derived. Theoretical analysis and Monte Carlo simulation show that the performance of the proposed multi-user cooperative NOMA system is better than that of the single-relay user cooperative system.

1 Introduction

Since the available frequency band of mobile communication is mainly below 3 GHz, 5G networks need to meet the requirements of low latency and massive connections under limited spectrum resources. Non-orthogonal multiple access (NOAM) is a important technology to solve the shortage of spectrum resources. Its basic principle is that the transmitter adopts power domain superposition coding, actively

M. Ma · Y. He (✉) · Y. Zhang · Y. Zhang · L. Zhou
Xiamen Key Laboratory of Mobile Multimedia Communications, National Huaqiao University, Xiamen 361021, Fujian, China
e-mail: yucheng.he@hqu.edu.cn

Y. He · L. Zhou
State Key Laboratory of Integrated Services Networks, Xidian University, Xi'an 710071, Shaanxi, China

L. C. Jain et al. (eds.), *Advances in Wireless Communications and Applications*, Smart Innovation, Systems and Technologies 299,
https://doi.org/10.1007/978-981-19-2255-8_3

introducing interference information, and the receiver adopts successive interference cancellation (SIC) to realize correct demodulation [1].

Due to shadow fading and path loss, the communication reliability of users far from the transmitter is poor. Relay cooperative NOMA network can improve the communication quality of far users. According to the NOMA superposition coding principle, strong user know the signal information of weak users and act as a Decode-and-Forward (DF) relay to assist in forwarding the signal information of weak users [2]. Under the statistical channel state information (CSI), the authors analyzed a communication system in which a single relay forwards the signals of multiple edge users over Nakagami-m fading channels, and the differences between DF relay and amplify-and-forward (AF) relay in improving communication performance was compared in [3]. In addition, device-to-device (D2D) technology allows two adjacent user equipment to communicate directly by reusing the spectrum resources of the cellular network, thereby reducing communication delay and improve spectrum utilization. So D2D relay cooperative communication based on NOMA has attracted extensive research scholars' attention. For D2D cooperative NOMA network, a power allocation strategy based on signal-to-noise ratio (SNR) was proposed in [4] to achieve the maximum capacity. Moreover, two different SIC decoding schemes were proposed in [5], namely the single signal decoding scheme and the maximum ratio combining (MRC) decoding scheme, respectively, and the authors compared and analyzed the influence of different decoding schemes on communication performance. In order to avoid the loss of communication performance caused by SIC error decoding, according to the decoding of D2D relay, three different relay forwarding strategies were studied in [6].

Due to the openness of RF signals, sensitive information with high security requirements is easily attacked by eavesdroppers during transmission. Therefore, the security of NOMA network must be considered. Relay cooperation can not only improve the communication reliability of cell-edge users, but also help legitimate nodes improve the physical layer security of the system. Physical layer security technology is based on Shannon theory. Its main idea is to protect the physical layer by using the inherent random characteristics of wireless channel [7]. At present, many researches on physical layer security are relay selection, cooperative beamforming, cooperative interference and so on. According to the different understanding of the eavesdropping channel, the corresponding relay selection strategy was proposed to combat the malicious eavesdropping of adaptive eavesdroppers in [8]. Considering the problem of energy limitation in communication, a AF relay collected energy from part of artificial interference signals, and forwarded the remaining artificial interference signals to eavesdroppers, so as to protect the legitimate signals [9]. In [10], the multi-antenna relay beamforming technology was used to design artificial interference signals to interfere with eavesdroppers without affecting the communication of legitimate users.

At present, there is little research on D2D cooperative NOMA scheme in the intelligent eavesdropping scenario. Furthermore, each user has different requirements for information security service experience. For example, sensitive information such as bank information and personal privacy information has high security requirements, but ordinary data information has low security requirements. So this paper proposes a multi-user cooperative NOMA scheme by using the differences of users' security service requirements, in order to further improve physical layer security, a user selection strategy was studied. We derive the expression on the ergodic secrecy sum capacity lower bound of the proposed scheme, and compare the performance differences between the scheme and single user cooperative NOMA scheme.

2 System Model

Figure 1 shows the system model of the proposed scheme, where a source node denoted S, a number N of DF relay user nodes denoted $\mathrm{R}_k (k = 1, 2, \ldots, N)$, two destination nodes denoted D_1 and D_2, and a eavesdropping node denoted E are considered. All nodes work in half-duplex mode, equipped with a single antenna. Moreover, E is equipped with a directional antenna. We assume that all channels are independent quasi-static Rayleigh fading channels. Denote $h_{\mathrm{S1}}, h_{\mathrm{SR}_k}, h_{\mathrm{R}_k\mathrm{E}}, h_{\mathrm{R}_k2}, h_{\mathrm{E1}}, h_{\mathrm{ER}_k}$ as the channel coefficients of links S - D_1, S - R_k, R_k - E, R_k - D_2, E - D_1, E - R_k respectively. According to [11], $h_{\mathrm{R}_k\mathrm{E}} = h_{\mathrm{ER}_k}$. Due to path loss and shadow fading, the links S - D_2 and S - E are not available. $h_{\mathrm{S1}}, h_{\mathrm{SR}_k}, h_{\mathrm{R}_k\mathrm{E}}, h_{\mathrm{R}_k2}, h_{\mathrm{E1}}, h_{\mathrm{ER}_k}$ are independent complex Gaussian random variables with mean denoted as λ_{S1}, λ_{SR_k}, $\lambda_{\mathrm{R}_k\mathrm{E}}$, λ_{R_k2}, λ_{E1}, $\lambda_{\mathrm{R}_k\mathrm{E}}$. Each node knows the CSI between itself and other nodes, and the CSI of the eavesdropping channel can be obtained by the method mentioned in [12]. Here E is an intelligent eavesdropper, which means that if E detects the signal transmitted by R_k, E will work in eavesdropping mode and directed eavesdropping R_k, otherwise it will switch to interference mode.

Fig.1 System model

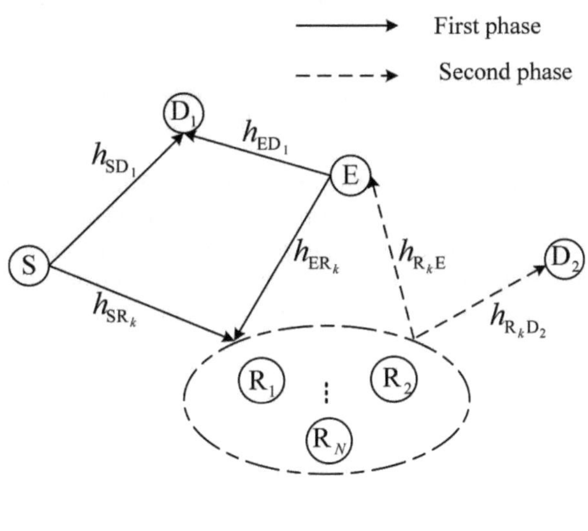

2.1 Design of Transmission Scheme

The entire transmission process comprises two phases. In the first phase, the source node S broadcasts the superimposed signal $\sqrt{a_1 P_S}x_1 + \sqrt{a_2 P_S}x_2$, where P_S is the transmit power of S, x_i represents the intended message for $D_i, i \in \{1, 2\}$. According to NOMA principles, the power allocation coefficient $a_1 < a_2$ and $a_1+a_2 = 1$. At the same time, the eavesdropper E executes interference mode and transmits interference signals $\sqrt{P_J}x_J$ to legitimate nodes. The signal received at D_1 and R are given by

$$y_1 = h_{S1}\left(\sqrt{a_1 P_S}x_1 + \sqrt{a_2 P_S}x_2\right) + h_{E1}\sqrt{P_J}x_J + n_1 \tag{1}$$

$$y_{R_k} = h_{SR_k}\left(\sqrt{a_1 P_S}x_1 + \sqrt{a_2 P_S}x_2\right) + h_{ER_k}\sqrt{P_J}x_J + n_{R_k} \tag{2}$$

where $n_1 \sim CN(0, \sigma_1)$ and $n_{R_k} \sim CN\left(0, \sigma_{R_k}\right)$ represent the additive white Gaussian noise AWGN at D_1 and R_k respectively. To abide with the principle of SIC, D_1 first decodes and removes the symbol x_2 by treating the symbol x_1 as noise, and then will decode symbol x_1. The received signal-to-interference-plus-noise ratio (SINR) for x_1 and x_2 at D_1 are given by

$$\gamma_{1\to x_2} = \frac{|h_{S1}|^2 a_2 P_S}{|h_{S1}|^2 a_1 P_S + |h_{E1}|^2 P_J + \sigma_1^2} \tag{3}$$

$$\gamma_{1\to x_1} = \frac{|h_{S1}|^2 a_1 P_S}{|h_{E1}|^2 P_J + \sigma_1^2} \tag{4}$$

Similarly, the received SINR for x_2 at R_k is expressed as

$$\gamma_{R_k\to x_2} = \frac{\left|h_{SR_k}\right|^2 a_2 P_S}{\left|h_{SR_k}\right|^2 a_1 P_S + \left|h_{ER_k}\right|^2 P_J + \sigma_{R_k}^2} \tag{5}$$

In the second phase, in order to alleviate the communication congestion and reduce the SINR of the eavesdropper, R_k forwards the new superimposed signal $\sqrt{a_1 P_R}x_R + \sqrt{a_2 P_R}x_2$ with the transmit power P_R. x_R is the normal data signal intended for D_2. Assuming that in the entire time slot, E can detect the signal forwarded by R, so it works in eavesdropping mode. The signal received at D_2 and E are given by

$$y_2 = h_{R_k 2}\left(\sqrt{a_1 P_R}x_R + \sqrt{a_2 P_R}x_2\right) + n_2 \tag{6}$$

$$y_E = h_{R_k E}\left(\sqrt{a_1 P_R}x_R + \sqrt{a_2 P_R}x_2\right) + n_E \tag{7}$$

where $n_2 \sim CN(0, \sigma_2)$ and $n_E \sim CN(0, \sigma_E)$ represent the AWGN at D_2 and E respectively. The received SINR for x_2 and x_R can be obtained as

$$\gamma_{2 \to x_2} = \frac{\left|h_{R_k 2}\right|^2 a_2 P_R}{\left|h_{R_k 2}\right|^2 a_1 P_R + \sigma_2^2} \tag{8}$$

$$\gamma_{2 \to x_R} = \frac{\left|h_{R_k 2}\right|^2 a_1 P_R}{\sigma_2^2} \tag{9}$$

Since E only decodes sensitive signals x_2 that have eavesdropping value, the received SINR at E is given by

$$\gamma_{E \to x_2} = \frac{\left|h_{R_k E}\right|^2 a_2 P_R}{\left|h_{R_k E}\right|^2 a_1 P_R + \sigma_E^2} \tag{10}$$

In order to reduce the eavesdropping quality of E, we propose a relay selection strategy that minimizes the channel gain of eavesdropping, and the mathematical expression is as follows

$$R_b = \arg \min_{k=1,2,\ldots,N} \left|h_{R_k E}\right|^2 \tag{11}$$

3 Analysis of Ergodic Secrecy Sum Capacity

Here we use ergodic secrecy sum capacity (ESSC) as the performance indicator to evaluate the effectiveness of the proposed scheme. For the convenience of writing, let $\sigma_i^2 = \sigma^2$, for $i \in \{1, 2, R_b, E\}$, $\gamma_S = \frac{P_S}{\sigma^2}$, $\gamma_J = \frac{P_J}{\sigma^2}$, $\gamma_R = \frac{P_R}{\sigma^2}$, $\beta_\varphi \triangleq \left|h_\varphi\right|^2$, for $\varphi \in \{S1, SR_b, R_bE, R_b2, E1, ER_b\}$.

3.1 Ergodic Secrecy Capacity of x_1

Ergodic secrecy capacity (ESC) of x_1 is defined as $\overline{C}_{ESC}^{x_1} \triangleq \mathbb{E}\left[\left\{C_1^{x_1} - C_E^{x_1}\right\}^+\right]$, where $C_1^{x_1}$ and $C_E^{x_1}$ is the achievable maximum instantaneous rate of x_1 at D_1 and E respectively. According to Jensen's inequality, the definition of the ESC lower bound of x_1 can be expressed as

$$\overline{C}_{ESC,lb}^{x_1} \triangleq \left\{\mathbb{E}\left[C_1^{x_1}\right] - \mathbb{E}\left[C_E^{x_1}\right]\right\}^+ = \left\{\overline{C}_1^{x_1} - \overline{C}_E^{x_1}\right\}^+ \tag{12}$$

Let $Q_1 = \gamma_{1\to x_1}$, the complementary cumulative distribution function (CCDF) of Q_1 is written as

$$\overline{F}_{Q_1}(q_1) = \Pr\left(\frac{\beta_{S1} a_1 P_S}{\beta_{E1} P_J + \sigma^2} > q_1\right) = \frac{\lambda_{S1} a_1 \gamma_S}{\lambda_{S1} a_1 \gamma_S + \lambda_{E1} \gamma_J q_1} e^{-\frac{q_1}{\lambda_{S1} a_1 \gamma_S}} \tag{13}$$

Using the equation in [5]

$$\int_0^\infty \log(1+x) f_X(x) dx = \frac{1}{\ln 2} \int_0^\infty \frac{1 - F(x)}{1+x} dx \tag{14}$$

and [13, Eq. (25.4)], we can finally obtain

$$\approx \frac{\pi^2}{8L_1 \ln 2} \sum_{i=1}^{L_1} \frac{\lambda_{S1} a_1 \gamma_S \sqrt{1-\alpha_i} \sec^2 \theta_i}{(\lambda_{S1} a_1 \gamma_S + \lambda_{E1} \gamma_J \tan \theta_i)(1 + \tan \theta_i)} e^{-\frac{\tan \theta_i}{\lambda_{S1} a_1 \gamma_S}} \tag{15}$$

where $\alpha_i = \cos \frac{(2i-1)\pi}{2L_1}$, $\theta_i = \frac{\pi}{4}\alpha_i + \frac{\pi}{4}$, L_1 denotes the number of the Gauss–Chebyshev quadrature approximation. Due to path loss and shadows, the eavesdropper E cannot eavesdrop on the superimposed signal broadcasted by S, thus $\overline{C}_E^{x_1} = 0$ ◦Substituting $\overline{C}_E^{x_1} = 0$ and (15) into (12), the expression of the ESC lower bound of x_1 can be obtained as

$$\overline{C}_{ESC,lb}^{x_1} \approx \left\{ \frac{\pi^2}{8L_1 \ln 2} \sum_{i=1}^{L_1} \frac{\lambda_{S1} a_1 \gamma_S \sqrt{1-\theta_i} \sec^2 \theta_i}{(\lambda_{S1} a_1 \gamma_S + \lambda_{E1} \gamma_J \tan \theta_i)(1 + \tan \theta_i)} e^{-\frac{\tan \theta_i}{\lambda_{S1} a_1 \gamma_S}} \right\}^+ \tag{16}$$

3.2 *Ergodic Secrecy Capacity of x_2*

Similar to x_1, using Jensen's inequality, the definition of the ESC lower bound of x_2 can be expressed as

$$\overline{C}_{ESC,lb}^{x_2} \triangleq \{\mathbb{E}[C_2^{x_2}] - \mathbb{E}[C_E^{x_2}]\}^+ = \left\{\overline{C}_2^{x_2} - \overline{C}_E^{x_2}\right\}^+ \tag{17}$$

where $C_2^{x_2}$ and $C_E^{x_2}$ is the achievable maximum instantaneous rate of x_2 at D_1 and E. Let $T_1 = \min\{\gamma_{1\to x_2}, \gamma_{R_k \to x_2}, \gamma_{2\to x_2}\}$, $T_2 = \gamma_{E\to x_2}$, $V_1 = \beta_{R_b E}$, where T_1 is the effective received SNR for x_2. Due to $|h_{R_1 E}|^2, |h_{R_2 E}|^2, \ldots, |h_{R_N E}|^2$ are an independent and identically distributed random variable, the probability density function (PDF) of V_1 is $f_{V_1}(v_1) = \frac{N}{\lambda_{RE}} e^{-\frac{N}{\lambda_{RE}} v_1}$. The achievable ESC of x_2 at D_2 and E can be written as

$$\overline{C}_2^{x_2} = \frac{1}{2}\int\limits_0^\infty \log(1+t_1) f_{T_1}(t_1)dt_1 = \frac{1}{2\ln 2}\int\limits_0^\infty \frac{1-F_{T_1}(t_1)}{1+t_1}dt_1 \tag{18}$$

$$\overline{C}_{\mathrm{E}}^{x_2} = \frac{1}{2}\int\limits_0^\infty \log(1+t_2) f_{T_2}(t_2)dt_2 = \frac{1}{2\ln 2}\int\limits_0^\infty \frac{1-F_{T_2}(t_2)}{1+t_2}dt_2 \tag{19}$$

In order to obtain the expression of $C_2^{x_2}$ and $C_{\mathrm{E}}^{x_2}$, we derive the CCDF of T_1 and T_2 as follows:

$$\begin{aligned}\overline{F}_{T_1}(t_1) &= \Pr\left(\min\left\{\gamma_{1\to x_2}, \gamma_{\mathrm{R}_k\to x_2}, \gamma_{2\to x_2}\right\} > t_1\right)\\ &= \frac{N}{\lambda_{\mathrm{E1}}\lambda_{\mathrm{ER}_b}} \mathrm{e}^{-\left(\frac{1}{\lambda_{\mathrm{S1}}\gamma_{\mathrm{S}}}+\frac{1}{\lambda_{\mathrm{SR}}\gamma_{\mathrm{S}}}+\frac{1}{\lambda_{\mathrm{R2}}\gamma_{\mathrm{R}}}\right)\frac{t_1}{(a_2-a_1t_1)}}\\ &\quad \int\limits_0^\infty \mathrm{e}^{-u\left(\frac{\gamma_{\mathrm{J}}t_1}{\lambda_{\mathrm{S1}}(a_2-a_1t_1)\gamma_{\mathrm{S}}}+\frac{1}{\lambda_{\mathrm{E1}}}+\frac{\gamma_{\mathrm{J}}t_1}{\lambda_{\mathrm{SR}}(a_2-a_1t_1)\gamma_{\mathrm{S}}}+\frac{N}{\lambda_{\mathrm{ER}_b}}\right)}du\\ &= \frac{\lambda_{\mathrm{S1}}\lambda_{\mathrm{SR}}\gamma_{\mathrm{S}}^2 N(a_2-a_1t_1)^2 \mathrm{e}^{-\left(\frac{1}{\lambda_{\mathrm{S1}}\gamma_{\mathrm{S}}}+\frac{1}{\lambda_{\mathrm{SR}}\gamma_{\mathrm{S}}}+\frac{1}{\lambda_{\mathrm{R2}}\gamma_{\mathrm{R}}}\right)\frac{t_1}{(a_2-a_1t_1)}}}{\left[\lambda_{\mathrm{S1}}\gamma_{\mathrm{S}}(a_2-a_1t_1)+\lambda_{\mathrm{E1}}\gamma_{\mathrm{J}}t_1\right]\left[\lambda_{\mathrm{SR}}\gamma_{\mathrm{S}}N(a_2-a_1t_1)+\lambda_{\mathrm{ER}}\gamma_{\mathrm{J}}t_1\right]}\end{aligned} \tag{20}$$

$$\overline{F}_{T_2}(t_2) = \Pr\left(\gamma_{\mathrm{E}\to x_2} > t_2\right) = \Pr\left(\frac{\beta_{\mathrm{R}_k\mathrm{E}}a_2P_{\mathrm{R}}}{\beta_{\mathrm{R}_k\mathrm{E}}a_1P_{\mathrm{R}}+\sigma_{\mathrm{E}}^2} > t_2\right) = \mathrm{e}^{-\frac{Nt_2}{\lambda_{\mathrm{RE}}\gamma_{\mathrm{R}}(a_2-a_1t_2)}} \tag{21}$$

Substituting (20) into (18), the achievable ESC of x_2 at D_2 is given by

$$\overline{C}_2^{x_2} = \frac{\pi}{2L_2\ln 2}\sum_{i=1}^{L_2}\frac{\lambda_{\mathrm{S1}}\lambda_{\mathrm{SR}}\gamma_{\mathrm{S}}^2a_1^2a_2N(1-x_i)^2\sqrt{1-y_i^2}\times \mathrm{e}^{-\left(\frac{1}{\lambda_{\mathrm{S1}}\gamma_{\mathrm{S}}}+\frac{1}{\lambda_{\mathrm{SR}}\gamma_{\mathrm{S}}}+\frac{1}{\lambda_{\mathrm{R2}}\gamma_{\mathrm{R}}}\right)\frac{(1+x_i)}{a_1(1-x_i)}}}{A(x_i)} \tag{22}$$

where L_2 denotes the number of the Gauss–Chebyshev quadrature approximation, $y_i = \cos\frac{(2i-1)\pi}{2L_2}$, $x_i = y_i$, and

$$\begin{aligned}A(x_i) &= (2a_1+a_2(1+x_i))(\lambda_{\mathrm{S1}}\gamma_{\mathrm{S}}a_1(1-x_i)+\lambda_{\mathrm{E1}}\gamma_{\mathrm{J}}(1+x_i))\\ &\quad\times(\lambda_{\mathrm{SR}}\gamma_{\mathrm{S}}Na_1(1-x_i)+\lambda_{\mathrm{ER}}\gamma_{\mathrm{J}}(1+x_i))\end{aligned} \tag{23}$$

Using (19), (21) and [14, Eq. (3.352.2)], the achievable ESC of x_2 at E is given by

$$\overline{C}_{\mathrm{E}}^{x_2} = \frac{1}{2\ln 2}\left[\mathrm{e}^{\frac{N}{\lambda_{\mathrm{RE}}\gamma_{\mathrm{R}}a_1}}\mathrm{Ei}\left(-\frac{N}{\lambda_{\mathrm{RE}}\gamma_{\mathrm{R}}a_1}\right) - \mathrm{e}^{\frac{N}{\lambda_{\mathrm{RE}}\gamma_{\mathrm{R}}}}\mathrm{Ei}\left(-\frac{N}{\lambda_{\mathrm{RE}}\gamma_{\mathrm{R}}}\right)\right] \tag{24}$$

where $\mathrm{Ei}(\cdot)$ is exponential integral function. Therefore, the expression of the ESC lower bound of x_2 can be finally obtained as

$$\overline{C}_{\mathrm{ESC,lb}}^{x_2} \approx \frac{1}{2\ln 2}\left\{\begin{array}{l} \frac{\pi}{L_2}\sum_{i=1}^{L_2} \frac{\lambda_{\mathrm{S1}}\lambda_{\mathrm{SR}}\gamma_{\mathrm{S}}^2 a_1^2 a_2 N(1-x_i)^2\sqrt{1-y_i^2}\times \mathrm{e}^{-\left(\frac{1}{\lambda_{\mathrm{S1}}\gamma_{\mathrm{S}}}+\frac{1}{\lambda_{\mathrm{SR}}\gamma_{\mathrm{S}}}+\frac{1}{\lambda_{\mathrm{R2}}\gamma_{\mathrm{R}}}\right)\frac{(1+x_i)}{a_1(1-x_i)}}}{A(x_i)} \\ -\left[\mathrm{e}^{\frac{N}{\lambda_{\mathrm{RE}}\gamma_{\mathrm{R}}a_1}}\mathrm{Ei}\left(-\frac{N}{\lambda_{\mathrm{RE}}\gamma_{\mathrm{R}}a_1}\right)-\mathrm{e}^{\frac{N}{\lambda_{\mathrm{RE}}\gamma_{\mathrm{R}}}}\mathrm{Ei}\left(-\frac{N}{\lambda_{\mathrm{RE}}\gamma_{\mathrm{R}}}\right)\right] \end{array}\right\}^{+} \tag{25}$$

3.3 Ergodic Secrecy Capacity of x_{R} and ESSC of System

Since x_{R} is a ordinary data signal, and E only intercepts valuable and sensitive signals, the ergodic secrecy capacity of x_{R} is its ergodic capacity. Let $V_2 = \gamma_{2\to x_{\mathrm{R}}}$, CCDF of V_2 is as follows

$$\overline{F}_{V_2}(v_2) = \Pr\left(\frac{\left|h_{\mathrm{R}_k 2}\right|^2 a_1 P_{\mathrm{R}}}{\sigma_2^2} > v_2\right) = \mathrm{e}^{-\frac{v_2}{\lambda_{\mathrm{R2}}\gamma_{\mathrm{R}}a_1}} \tag{26}$$

According to [14, Eq. (3.352.4)], the ergodic secrecy capacity of x_{R} can be written as

$$\begin{aligned}\overline{C}_{\mathrm{ESC}}^{x_{\mathrm{R}}} &= \frac{1}{2}\int_0^\infty \log(1+v_1) f_{V_1}(v_1)dv_1 = \frac{1}{2\ln 2}\int_0^\infty \frac{1-F_{V_1}(v_1)}{1+v_1}dv_1 \\ &= -\frac{1}{2\ln 2}\mathrm{e}^{\frac{1}{\lambda_{\mathrm{R2}}\gamma_{\mathrm{R}}a_1}}\mathrm{Ei}\left(-\frac{1}{\lambda_{\mathrm{R2}}\gamma_{\mathrm{R}}a_1}\right)\end{aligned} \tag{27}$$

So the ESSC of system is approximated as

$$\begin{aligned}\overline{C}_{\mathrm{ESSC,lb}} &= \overline{C}_{\mathrm{ESC,lb}}^{x_{\mathrm{R}}} + \overline{C}_{\mathrm{ESC,lb}}^{x_1} + \overline{C}_{\mathrm{ESC,lb}}^{x_2} \approx -\frac{1}{2\ln 2}\mathrm{e}^{\frac{1}{\lambda_{\mathrm{R2}}\gamma_{\mathrm{R}}a_1}}\mathrm{Ei}\left(-\frac{1}{\lambda_{\mathrm{R2}}\gamma_{\mathrm{R}}a_1}\right) \\ &+\left\{\frac{\pi^2}{8L_1\ln 2}\sum_{i=1}^{L_1}\frac{\lambda_{\mathrm{S1}}a_1\gamma_{\mathrm{S}}\sqrt{1-\theta_i}\sec^2\theta_i}{(\lambda_{\mathrm{S1}}a_1\gamma_{\mathrm{S}}+\lambda_{\mathrm{E1}}\gamma_{\mathrm{J}}\tan\theta_i)(1+\tan\theta_i)}\mathrm{e}^{-\frac{\tan\theta_i}{\lambda_{\mathrm{S1}}a_1\gamma_{\mathrm{S}}}}\right\}^{+} \\ &+\frac{1}{2\ln 2}\left\{\begin{array}{l} \frac{\pi}{L_2}\sum_{i=1}^{L_2} \frac{\lambda_{\mathrm{S1}}\lambda_{\mathrm{SR}}\gamma_{\mathrm{S}}^2 a_1^2 a_2 N(1-x_i)^2\sqrt{1-y_i^2}\times \mathrm{e}^{-\left(\frac{1}{\lambda_{\mathrm{S1}}\gamma_{\mathrm{S}}}+\frac{1}{\lambda_{\mathrm{SR}}\gamma_{\mathrm{S}}}+\frac{1}{\lambda_{\mathrm{R2}}\gamma_{\mathrm{R}}}\right)\frac{(1+x_i)}{a_1(1-x_i)}}}{A(x_i)} \\ -\left[\mathrm{e}^{\frac{N}{\lambda_{\mathrm{RE}}\gamma_{\mathrm{R}}a_1}}\mathrm{Ei}\left(-\frac{N}{\lambda_{\mathrm{RE}}\gamma_{\mathrm{R}}a_1}\right)-\mathrm{e}^{\frac{N}{\lambda_{\mathrm{RE}}\gamma_{\mathrm{R}}}}\mathrm{Ei}\left(-\frac{N}{\lambda_{\mathrm{RE}}\gamma_{\mathrm{R}}}\right)\right] \end{array}\right\}^{+}\end{aligned} \tag{28}$$

4 Numerical Analysis

In this section, we will evaluate the performance of our proposed D2D-NOMA system in terms of ESSC under the user selection scheme through simulation. The relevant parameter settings in the simulation are as follows: $L_1 = L_2 = 100$, legal channel parameters $\lambda_{S1} = \lambda_{SR} = 1$, $\lambda_{R2} = 2$, in order to avoid being discovered, the eavesdropper is far away from the legal nodes, so $\lambda_{RE} = \lambda_{E1} = 0.1$. If there is no special instructions, we set $\gamma_S = \gamma_R = 30\,\text{dB}$, $a_1 = 0.2$, $a_2 = 0.8$ and $N = 10$. Similarly, due to the need for concealment, $\gamma_J = 5\,\text{dB}$. The number of Monte Carlo simulations is 1,000,000.

For comparison, we consider the following benchmark schemes in the simulation.

(1) Benchmark scheme 1 is a multi-relay user cooperative NOMA scheme, in which the number of relay users is $N = 10$. In the first time slot, S broadcast superimposed signal $\sqrt{a_1 P_S} x_1 + \sqrt{a_2 P_S} x_2$, at the same time the eavesdropper transmits interference signals $\sqrt{P_J} x_J$. In the second time slot, the selected relay forward the signal x_2 with the transmit power P_R, and E switches to eavesdropping mode.

(2) Benchmark scheme 2 is a single-relay user cooperative NOMA scheme, in which the number of relay users is $N = 1$. In first time slot, S broadcast superimposed signal $\sqrt{a_1 P_S} x_1 + \sqrt{a_2 P_S} x_2$, and the eavesdropper transmits interference signals $\sqrt{P_J} x_J$. In the second time slot, relay user forwards signal $\sqrt{a_1 P_R} x_R + \sqrt{a_2 P_J} x_2$, and E switches to eavesdropping mode.

Figure 2 shows the relationship between the number of relay users and the system ESSC under different power allocation factors. As shown in Fig. 1, as the number of

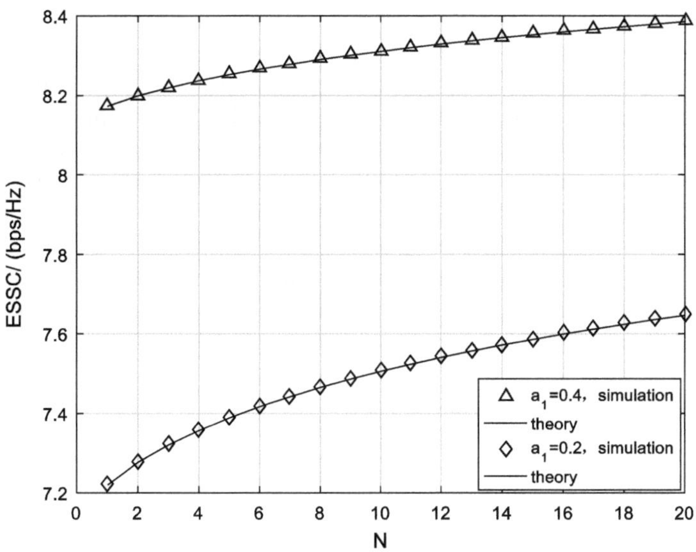

Fig. 2 The effect of N on system ESSC with different power allocation factors

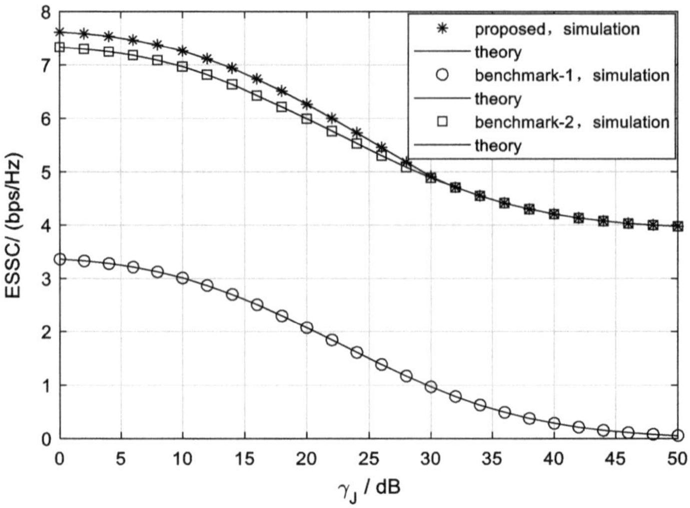

Fig. 3 The effect of γ_J on system ESSC with different transmission schemes

relay users increases, the ESSC of the system increases monotonically, and compared to $a_1 = 0.4$, the number of relay users has a greater impact on the system ESSC when $a_1 = 0.2$. It can also be observed from the figure, secrecy performance of the system when $a_1 = 0.4$ is better than that of the system when $a_1 = 0.2$. This is because there is more power allocated to x_1 and x_R when $a_1 = 0.4$, which is beneficial for D_1 to resist the interference attack of E, furthermore the eavesdropper does not eavesdrop on the ordinary data signals x_R.

Figure 3 simulates the influence of the eavesdropper's transmitted SINR on the system ESSC. The simulation shows that the security performance of the proposed scheme is better than that of Benchmark scheme 1 and Benchmark scheme 2, and the performance gap between the proposed scheme and Benchmark scheme 1 is more obvious. It implies that it is of great significance to design a secure communication scheme based on the differences in user experience requirements for information security services. In addition, it can also be observed that the ESSC of the proposed scheme and Benchmark scheme 2 tend to a positive definite value in the high SIR, this is because the transmission of x_R occurs in the second time slot, so it will not be affected by the interference of E. By comparing the curves of Monte Carlo simulation and theoretical analysis, it can be found that the two basically overlap, reflecting the accuracy of the numerical calculation.

5 Conclusions

This paper combines user selection technology with D2D technology to design a NOMA communication scheme against intelligent eavesdropping, and derives the expression of the ESSC lower bound of the proposed scheme. Theoretical derivation and simulation show that the proposed scheme is better than the two Benchmark schemes. The communication system can take advantage of the differences in information security service experience requirements of each user to alleviate communication congestion and improve the overall security performance. In addition, by increasing the number of relay users, the security of information transmission can be further ensured.

References

1. Wang, G., Xu, X., Zhou, R., Zhang, R.: Power domain non-orthogonal multiple access technology for wireless communication. Radio Commun. Technol. **45**(4), 329–336 (2019)
2. Ding, Z., Peng, M., Poor, H.: Cooperative non-orthogonal multiple access in 5G systems. IEEE Commun. Lett. **19**(8), 1462–1465 (2015)
3. Wan, D., Wen, M., Ji, F., Liu, Y., Huang, Y.: Cooperative NOMA systems with partial channel state information over Nakagami-m fading channels. IEEE Trans. Commun. **66**(3), 947–958 (2018)
4. Kim, J., Lee, I., Lee, J.: Capacity scaling for D2D aided cooperative relaying systems using NOMA. IEEE Commun. Lett. **7**(1), 42–45 (2018)
5. Ji, Y., Wen, M., Padidar, P., Duan, W., Li, J., Cheng, N., Ho, P.: Spectral efficiency enhanced cooperative device-to-device systems with NOMA. IEEE Trans. Intell. Transp. Syst. **22**(7), 4040–4050 (2021)
6. Duan, W., Ji, Y., Hou, J., Zhuo, B., Wen, M., Zhang, G.: Partial-DF full-duplex D2D-NOMA systems for IoT with/without an eavesdropper. IEEE Internet Things J. **8**(8), 6154–6166 (2021)
7. Zou, Y., Zhu, J., Wang, X., Hanzo, L.: A survey on wireless security: technical challenges, recent advances, and future trends. Proc. IEEE **104**(9), 1727–1756 (2016)
8. Yang, L., Chen, J., Jiang, H., Vorobyov, S., Zhang, H.: Optimal relay selection for secure cooperative communications with an adaptive eavesdropper. IEEE Trans. Wireless Commun. **16**(1), 26–42 (2017)
9. Lee, K., Hong, J., Choi, H., Levorato, M.: Adaptive wireless-powered relaying schemes with cooperative jamming for two-hop secure communication. IEEE Internet Things J. **5**(4), 2793–2803 (2018)
10. Cao, Y., Zhao, N., Pan, G., Chen, Y., Fan, L., Jin, M., Alouini, M.: Secrecy analysis for cooperative NOMA networks with multi-antenna full-duplex relay. IEEE Trans. Commun. **67**(8), 5574–5587 (2019)
11. Bletsas, A., Shin, H., Win, M.: Cooperative communications with outage-optimal opportunistic relaying. IEEE Trans. Wireless Commun. **6**(9), 3450–3460 (2007)
12. Shim, K., Do, T., Nguyen, T., Costa, D., An, B.: Enhancing PHY-security of FD-enabled NOMA systems using Jamming and user selection: performance analysis and DNN evaluation. IEEE Internet Things J. https://doi.org/10.1109/JIOT.2021.3080425
13. Abramowitz, M., Stegun, I.: Handbook Mathematical Functions with Formulas, Graphs, Mathematical Tables. New York, NY , USA, Dover (1972)
14. Gradshteyn, I., Ryzhik, I.: Table of Integrals, Series and Products, 7th edn. NY, USA, Academic, New York (2007)

Performance Analysis and Relay Selection of D2D Aided Cooperative NOMA System

Shuting Wu, Yucheng He, Yu Zhang, Yan Zhang, and Lin Zhou

Abstract This paper investigates a cooperative non-orthogonal multiple access (NOMA) network, where cell center users (CU) can directly communicate with base stations (BS), while cell edge users (EU) needs to have half-duplex relay to assist in transmission. Specially, a D2D communication link from relay to CU is designed, which fully exploits the spectral resource to transmit a new signal. In addition, a two-stage relay selection strategy (TSRS) is proposed, which maximizes the probability of CU's successful decoding under the premise of ensuring reliable reception of the EU. To evaluate the proposed D2D aided cooperative relaying using NOMA (DC-NOMA) scheme, exact outage probabilities of each user data are derived and confirmed by Monte-Carlo computer simulations. By analyzing the outage probability, the performance of the proposed DC-NOMA using the TSRS outperforms that of the partial relay selection scheme (PRSS). In particular, the increase in the number of relays can effectively improve the outage performance of the proposed DC-NOMA using the TSRS network.

1 Introduction

With the explosive growth of Internet of Things (IoT) devices and user services, future wireless communications network will face major challenges in high spectral efficiency, low latency and massive connectivity [1]. Against this background, a key technology for next-generation cellular communications is NOMA since it allows multiple users to allocate different power domains by sharing the same frequency/time/code resources [2, 3]. Many researchers are dedicated to studying

S. Wu · Y. He (✉) · Y. Zhang · Y. Zhang · L. Zhou
Xiamen Key Laboratory of Mobile Multimedia Communications, National Huaqiao University, Xiamen 361021, Fujian, China
e-mail: yucheng.he@hqu.edu.cn

Y. He · L. Zhou
State Key Laboratory of Integrated Services Networks, Xidian University, Xi'an 710071, Shaanxi, China

L. C. Jain et al. (eds.), *Advances in Wireless Communications and Applications*, Smart Innovation, Systems and Technologies 299,
https://doi.org/10.1007/978-981-19-2255-8_4

NOMA integrated with existing advanced schemes such as cooperative relaying system (CRS) [4], device-to-device (D2D) [5] and cognitive radio (CR) [6].

Relay technology is widely used in modern wireless communication networks because of its advantages in effectively expanding network coverage and improving communication reliability. Reference [7] studies a downlink full-duplex relay system using NOMA and considers the influence of self-interference factor on the system outage probability. D2D communication has now attracted intensive attention due to it can improve the local spectral efficiency. The key idea of D2D is to allow direct communication between two adjacent devices without a base station (BS) involved [8]. The work in [9] proposes cooperative D2D communication with multiple D2D pairs, and optimal power allocation is obtained. A D2D-aided C-NOMA (DC-NOMA) was presented in [10] to enhance the outage performance of the EU link using NOMA. An appropriate power allocation factor is selected, the ergodic sum capacities of DC-NOMA becomes better than C-NOMA and C-OMA [11].

Considering the multi-user relay system, the diamond-like relay selection model is studied [12]. On the basis of the max–min scheme, performance analysis of two relay modes are carried out: amplify-and-forward (AF) and decode-and-forward (DF). The simulation results show that DF relay performs better than AF relay. In addition, the relay selection scheme can further improve the performance of DC-NOMA. In the cooperative DF relay system using NOMA, PRSS is used in literature [13] and system outage performances are improved. In DC-NOMA system, this letter proposes a TSRS and PRSS, and the outage probability of closed-form expression are derived for each user data information.

2 System Model

It was illustrated in Fig. 1, a DC-NOMA network consisting of a base station (BS), a EU, a CU, and N relays $R_n(n = 1, ..., N)$ is considered in this paper. Due to the path loss, there is no direct link between BS and EU. Thus, BS needs send

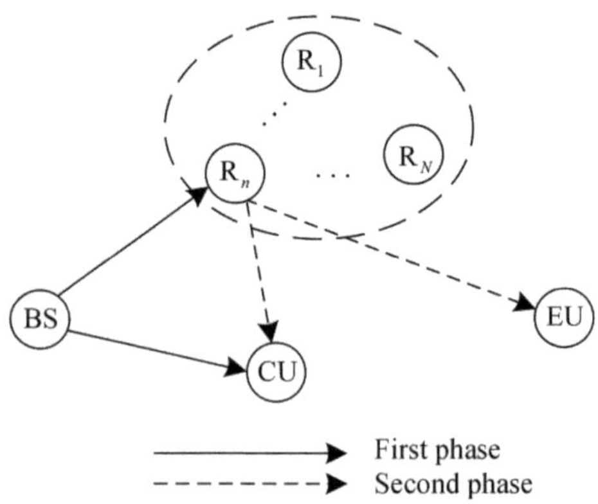

Fig. 1 A DC-NOMA system model

information to EU helped by N DF relays. In return, the relay can directly send itself signal to CU via D2D communication. All channels are Rayleigh fading, the channel coefficients between BS $\rightarrow$ R_n, BS $\rightarrow$ CU, R_n $\rightarrow$ EU and R_n $\rightarrow$ CU are represented as h_{sn}, h_{sc}, h_{ne} and h_{nc}, variances are expressed as λ_{ij} and zero mean, where $ij \in \{sn, sc, ne, nc\}$. Moreover, noise model of receiver is complex additive white Gaussian noise (AWGN).

In the first phase, BS sends a superimposed signal $\sqrt{P_s a_1} x_e + \sqrt{P_s a_2} x_c$, where x_e and x_c denote the signal required by EU and CU, respectively. P_s denotes the total transmit power of BS, and the a_i denotes the power allocation coefficient, satisfied $a_2 < a_1$, where $a_1 + a_2 = 1$. Therefore, the received signals at CU and the relay are respectively given by

$$y_{s\rightarrow n} = h_{sn}(\sqrt{P_s a_1} x_e + \sqrt{P_s a_2} x_c) + n_n \tag{1}$$

$$y_{s\rightarrow c} = h_{sc}(\sqrt{P_s a_1} x_e + \sqrt{P_s a_2} x_c) + n_c \tag{2}$$

where n_c and n_n denote the additive white Gaussian noise with σ^2 variance at CU and the relay. According to NOMA principle, the decoding signal-to-interference-plus-noise ratio (SINR) can be expressed by

$$\gamma_{sc,x_e} = \frac{a_1 |h_{sc}|^2}{a_2 |h_{sc}|^2 + 1/\rho} \tag{3}$$

where $\rho = P_s/\sigma^2$ is the transmit signal-to-noise ratio (SNR). Perform SIC on the CU and SNR with signal x_c written as

$$\gamma_{sc,x_c} = \rho |h_{sc}|^2 a_2 \tag{4}$$

Besides, the side information obtained by the CU through decoding is used.

In the second phase, when the relay forwards the decoded signal, it also sends its own signal to the D2D receiver (i.e., CU). Suppose that an optimal relay R_n is selected from N relays, this part will be discussed in Sect. 3. The DC-NOMA allows R_n applies the superposition coding technique to combine two independent signals $\sqrt{P_d b_1} x_e + \sqrt{P_d b_2} x_d$, where x_e is intended message to EU and x_d is a D2D signal intended to CU. What's more, b_1 and b_2 are the power allocation factor, assumed as $b_1 = a_1, b_2 = a_2$ and P_d denotes the transmitted power of R_n, where $P_d = \eta P_s$. Thus, the received signals at EU and CU are represented by

$$y_{n\rightarrow e} = h_{ne}(\sqrt{P_d b_1} x_e + \sqrt{P_d b_2} x_d) + n_e \tag{5}$$

$$y_{n\to c} = h_{nc}\left(\sqrt{P_d b_1}x_e + \sqrt{P_d b_2}x_d\right) + n_c \tag{6}$$

where n_e and n_c denote the AWGN at CU and the relay, respectively.

After evaluating the channel information h_{nc}, use the side information x_e obtained in SIC to eliminate the interference signal $h_{nc}\sqrt{P_d b_1}x_e$ from $y_{n\to c}$. Therefore, the CU can decode the signal x_d without interference, and its SINR can be indicated as:

$$\gamma_{nc,x_d} = \rho_d |h_{nc}|^2 b_2 \tag{7}$$

3 Relay Selection Scheme

3.1 Two-Stage Relay Selection

This relay selection scheme is usually divided into two stages. The TSRS scheme proposed in this paper selects the relay that maximizes the quality of data transmitted by the CU under the premise of ensuring the reliable reception of the signal by the EU. In the first stage, in order to ensure the reliable transmission of EU, a relay set of successfully decoded signals is established:

$$S_r = \left\{n : 1 \le n \le N, \gamma_{sn,x_e} \ge \gamma_E^{th}, \gamma_{ne,x_e} \ge \gamma_E^{th}\right\} \tag{8}$$

where $\gamma_E^{th} \triangleq 2^{2R_e} - 1$, and R_e is the target threshold for signal x_e.

It is worth noting that, whether x_c can be successfully decoded only related to the BS $\to$ CU link. Therefore, the relay selection scheme will not enhance the outage performance of x_c. In other words, it is only necessary to consider maximizing the probability of successful decoding x_d in the second stage. Select the relay n^* with the best $\mathrm{R}_n \to$ CU link channel state information in the relay subset:

$$n^* = \arg\max_n \left\{|h_{nc}|^2, n \in S_r\right\} \tag{9}$$

Next, the outage probability of the TSRS is derived to further analyze the system performance. According to formula (8), it can be seen that EU failed to decode successfully x_e, only when $|S_r| = 0$. Thus, the outage probability (OP) on the EU can be evinced as:

$$
\begin{aligned}
\mathrm{P}_{x_e}^{out,\mathrm{I}} &= \mathrm{P}[|S_r| = 0] \\
&= \prod_{n=1}^{N} \left(1 - \mathrm{P}\left[\gamma_{sn,x_e} \geq \gamma_E^{th}, \gamma_{ne,x_e} \geq \gamma_E^{th}\right]\right) \\
&= \prod_{n=1}^{N} \left(1 - \mathrm{P}\left[|h_{sn}|^2 \geq \frac{\Gamma_E^{th}}{\rho}\right]\mathrm{P}\left[|h_{ne}|^2 \geq \frac{\Gamma_E^{th}}{\rho_d}\right]\right) \\
&= \left(1 - e^{-\frac{\varphi_n}{\rho}}\right)^N
\end{aligned} \tag{10}
$$

where $\Gamma_E^{th} \triangleq \gamma_E^{th}/(a_1 - \gamma_E^{th} a_2)$ satisfied as $a_1 - \gamma_E^{th} a_2 > 0$, and $\varphi_n \triangleq \frac{\Gamma_E^{th}}{\lambda_{sn}} + \frac{\gamma_E^{th}}{\lambda_{ne}\eta}$ $(\forall n \in N)$.

Since an outage event occurs when CU cannot decode successfully x_e or x_c, the OP can be derived as

$$
\begin{aligned}
\mathrm{P}_{x_c}^{out,\mathrm{I}} &= \mathrm{P}\left[\gamma_{sc,x_e} < \gamma_E^{th} \cup \gamma_{sc,x_c} < \gamma_C^{th}\right] \\
&= 1 - \mathrm{P}\left[\gamma_{sc,x_e} \geq \gamma_E^{th} \cap \gamma_{sc,x_c} \geq \gamma_C^{th}\right] \\
&= 1 - \mathrm{P}\left[|h_{sc}|^2 \geq \frac{\Gamma_E^{th}}{\rho} \cap |h_{sc}|^2 \geq \frac{\gamma_C^{th}}{\rho a_2}\right] \\
&= 1 - e^{-\frac{1}{\rho\lambda_{sc}} \max\left\{\Gamma_E^{th}, \frac{\gamma_C^{th}}{a_2}\right\}}
\end{aligned} \tag{11}
$$

An outage event occurs at CU when: (i) $|S_r| = 0$, no relay is selected to send D2D signal x_d; and (ii) $|S_r| > 0$, CU cannot successfully decode x_e or x_d. Hence, the outage probability of x_d at CU can be formulated as

$$
\begin{aligned}
\mathrm{P}_{x_d}^{out} &= \mathrm{P}[|S_r| = 0] + \mathrm{P}\left[|S_r| > 0, \gamma_{sc,x_e} < \gamma_E^{th} \cup \gamma_{n^*c,x_d} < \gamma_D^{th}\right] \\
&= \mathrm{P}[|S_r| = 0] + \sum_{l=1}^{N} \mathrm{P}[|S_r| = l] \\
&\mathrm{P}\left[\gamma_{sc,x_e} < \gamma_E^{th} \cup \gamma_{n^*c,x_d} < \gamma_D^{th} \,|\, |S_r| = l\right]
\end{aligned} \tag{12}
$$

Since the channel gain satisfied the exponential distribution, according to the knowledge of probability theory, formula (12) can be divided into the following parts:

$$
\begin{aligned}
\mathrm{P}[|S_r| = l] &= \binom{N}{l} \prod_{n=1}^{N-l} (1 - \mathrm{P}[\gamma_{sn,x_e} \geq \gamma_E^{th}, \gamma_{ne,x_e} \geq \gamma_E^{th}]) \\
&\times \prod_{n=N-l+1}^{N} \mathrm{P}[\gamma_{sn,x_e} \geq \gamma_E^{th}, \gamma_{ne,x_e} \geq \gamma_E^{th}] \\
&= \binom{N}{l} \left[1 - e^{-\frac{\varphi_n}{\rho}}\right]^{N-l} e^{-\frac{\varphi_n}{\rho} l}
\end{aligned}
\tag{13}
$$

$$
\mathrm{P}[(\gamma_{sc,x_e} < \gamma_E^{th} \cup \gamma_{n^*c,x_d} < \gamma_D^{th}) || S_r| = l] = \left[1 - e^{-\frac{\Gamma_E^{th}}{\rho \lambda_{sc}}} e^{-\frac{\gamma_D^{th}}{\rho_d a_2 \lambda_{nc}}}\right]^l \tag{14}
$$

In summary, the closed-form expression for the outage probability of CU can be obtained as

$$
\begin{aligned}
\mathrm{P}_{x_d}^{out,\mathrm{I}} &= \mathrm{P}[|S_r| = 0] + \mathrm{P}[|S_r| > 0, \gamma_{sc,x_e} < \gamma_E^{th} \cup \gamma_{n^*c,x_d} < \gamma_D^{th}] \\
&= \mathrm{P}[|S_r| = 0] + \sum_{l=1}^{N} \mathrm{P}[|S_r| = l] \mathrm{P}[(\gamma_{sc,x_e} < \gamma_E^{th} \cup \gamma_{n^*c,x_d} < \gamma_D^{th}) || S_r| = l] \\
&= (1 - e^{-\frac{\varphi_n}{\rho}})^N + \sum_{l=1}^{N} \binom{N}{l} \left[1 - e^{-\frac{\varphi_n}{\rho}}\right]^{N-l} e^{-\frac{\varphi_n}{\rho} l} \left[1 - e^{-\frac{\Gamma_E^{th}}{\rho \lambda_{sc}}} e^{-\frac{\gamma_D^{th}}{\rho_d a_2 \lambda_{nc}}}\right]^l
\end{aligned}
\tag{15}
$$

3.2 *Partial Relay Selection*

The PRSS selects the best relay based on the channel quality of BS to the relay link, which can be expressed as

$$
n^* = \arg\max_{n} \{|h_{sn}|^2, n \in N\} \tag{16}
$$

According to the principle of relay selection, the probability density function of $|h_{sn}|^2$ can be defined as

$$
f_{|h_{sn}|^2}(x) = \sum_{n=1}^{N} \binom{N}{n} (-1)^{n-1} \frac{n}{\lambda_{sn}} e^{-\frac{n}{\lambda_{sn}} x} \tag{17}
$$

Hence, the outage probability of x_e at EU can be formulated as

$$\begin{aligned}
P_{x_e}^{out,II} &= 1 - P\left[\gamma_{sn,x_e} \geq \gamma_E^{th}, \gamma_{ne,x_e} \geq \gamma_E^{th}\right] \\
&= 1 - P\left[|h_{sn}|^2 \geq \frac{\Gamma_E^{th}}{\rho}\right] P\left[|h_{ne}|^2 \geq \frac{\Gamma_E^{th}}{\rho_d}\right] \\
&= 1 - \sum_{n=1}^{N} \binom{N}{n} (-1)^{n+1} e^{-\frac{n\Gamma_E^{th}}{\rho\lambda_{sn}}} e^{-\frac{\Gamma_E^{th}}{\rho_d\lambda_{ne}}}
\end{aligned} \tag{18}$$

The OP of x_d at CU can be derived by

$$\begin{aligned}
P_{x_d}^{out,II} &= 1 - P\left[\gamma_{sn,x_e} \geq \gamma_E^{th} \cap \gamma_{nc,x_d} \geq \gamma_D^{th} \cap \gamma_{sc,x_e} \geq \gamma_E^{th}\right] \\
&= 1 - P\left[|h_{sn}|^2 \geq \frac{\Gamma_E^{th}}{\rho} \cap |h_{nc}|^2 \geq \frac{\gamma_D^{th}}{\rho_d b_2} \cap |h_{sc}|^2 \geq \frac{\Gamma_E^{th}}{\rho}\right] \\
&= 1 - \left[\sum_{n=1}^{N} \binom{N}{n} (-1)^{n+1} e^{-\frac{n\Gamma_E^{th}}{\rho\lambda_{sn}}} \times e^{-\frac{\gamma_D^{th}}{\rho_d b_2 \lambda_{nc}}} \times e^{-\frac{\Gamma_E^{th}}{\rho\lambda_{sc}}}\right]
\end{aligned} \tag{19}$$

Furthermore, the outage probability of x_c has nothing to do with the relay selection, so $P_{x_c}^{out,II} = P_{x_c}^{out,I}$.

4 Performance Analysis

To confirm and compare the outage performance analysis, numerical results are provided in this section. Aiming at the proposed TSRS DC-NOMA network, Monte-Carlo computer simulation is carried out in a Rayleigh fading channel. Unless otherwise specified, the proposed system simulation parameters are set by default as: normalized distance $d_{sc} = d_{sn} = 0.6$, $d_{ne} = 0.4$, $d_{nc} = 0.25$.We consider the fixed power allocation factors, $a_1 = b_1 = 0.65$ and $a_2 = b_2 = 0.35$. Besides, the pass loss exponent are set as $a_{sc} = a_{sn} = a_{ne} = a_{nc} = 3$, and $\eta = 0.5$. Without loss of generality, $R_e = 0.7$bit/s, $R_c = 0.4$bit/s, and $R_d = 0.4$bit/s denote the target rate.

Figures 2 and 3 show the outage probability of x_e and x_d under the PRSS and TSRS relay selection schemes versus SNR, respectively. It can be clearly observed that analytical curves match with simulation results. It is worth noting that compared to the PRSS scheme, the proposed TSRS scheme can significantly improve the outage performance of x_e and x_d. For the x_e, the PRSS scheme only selects the relay with the best BS to the relay link quality, and only guarantees the outage probability of successful decoding x_e at the relay. However, the proposed TSRS in this paper is based on the premise of ensuring successful decoding on the relay and EU. For the signal x_d, the reason is that the TSRS scheme selects the relay with the best $R_n \rightarrow$ CU link quality under guarantee the reliable reception of the relay, which significantly reduces the OP of x_d.

Figure 4 illustrates how the number of relays affect outage probability of TSRS and PRSS scheme. Observe that the OP of PRSS scheme remains unchanged when

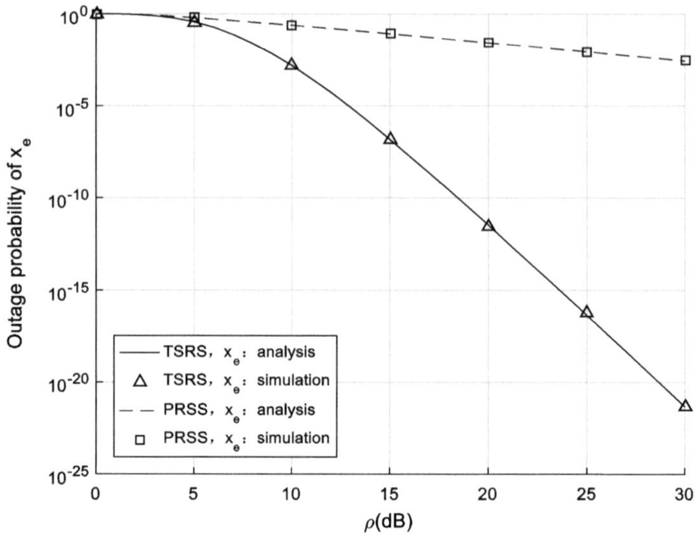

Fig. 2 Outage probability x_e versus of ρ

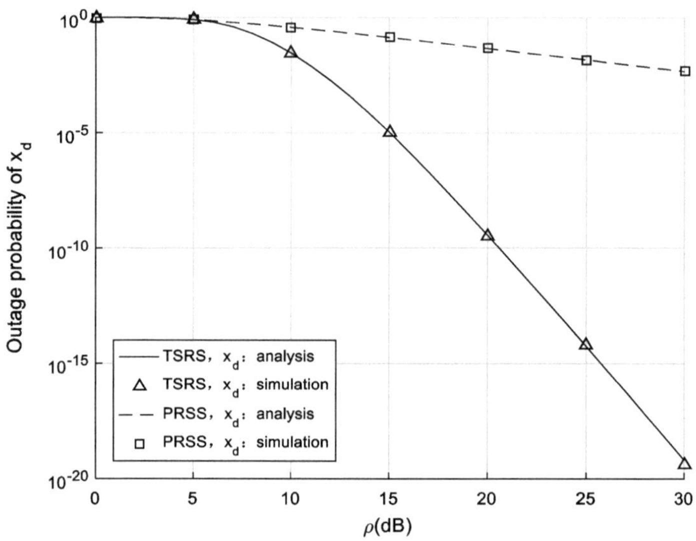

Fig. 3 Outage probability x_d versus of ρ

the relays reaches a certain value. Under the TSRS strategy, as the number of relays is increasing, the OP declines linearly. Noted that the superiority of TSRS scheme is apparent and the outage probability can be effectively reduced with the number of relays increasing. From a practical point of view, it is important to consider

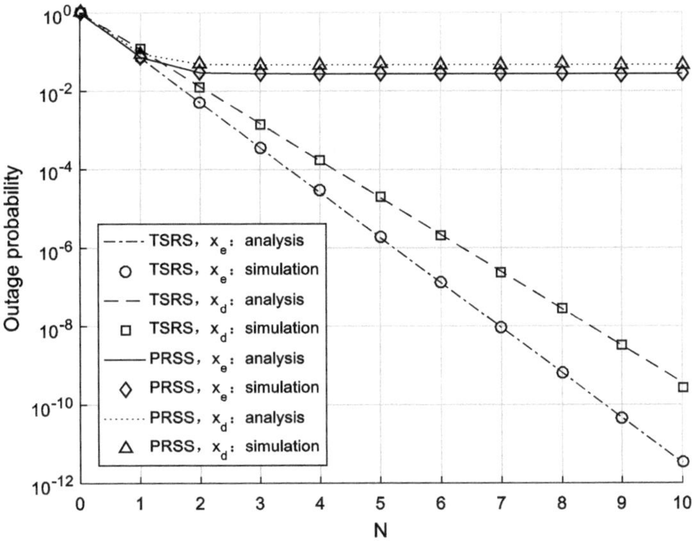

Fig. 4 Outage probability x_e and x_d versus of N

multiple relays scenarios. Furthermore, the TSRS scheme is especially significant for improving the outage performance of EU with poor channel quality.

5 Conclusion

In this paper, D2D communication is applied in the cooperative relaying NOMA network. In the future, the DC-NOMA can be exploited in the nearly local communication of the wireless communication network. Based on the DC-NOMA system, we have proposed a TSRS multiple relays selection scheme. The outage probabilities have been derived and the derivation results are fully consistent with simulation results. It was also shown that, as compared to PRSS scheme, the proposed TSRS scheme can effectively improve the outage performance of the DC-NOMA system. Noted that the increase in the number of relays greatly reduced the signal outage probability, so it can meet the application of large-scale relay scenarios in massive connections in the future.

References

1. Chettri, L., Bera, R.: A comprehensive survey on internet of things (IoT) toward 5G wireless systems. IEEE Internet Things J. **7**(1), 16–32 (2020)
2. Islam, S.M.R., Avazov, N., Dobre, O.A., Kwak, K.: Power-domain non-orthogonal multiple

access (NOMA) in 5G systems: potentials and challenges. IEEE Commun. Surv. Tutor. **19**(2), 721–742 (2017)
3. Li, X., Zhao, M., Liu, Y., Li, L., Nallanathan, A.: Secrecy analysis of ambient backscatter NOMA systems under I/Q imbalance. IEEE Trans. Veh. Technol. **69**(10), 12286–12290 (2020)
4. Yue, X., Liu, Y., Kang, S., Nallanathan, A., Ding, Z.: Exploiting full/half-duplex user relaying in NOMA systems. IEEE Trans. Commun. **66**(2), 560–575 (2018)
5. Ji, Y., Duan, W., Wen, M., Padidar, P., Li, J., Cheng, N., Ho, P.: Spectral efficiency enhanced cooperative device-to-device systems with NOMA. IEEE Trans. Intell. Transport. Syst. **22**(7), 4040–4050 (2021)
6. Lv, L., Chen, J., Ni, Q., Ding, Z., Jiang, H.: Cognitive non-orthogonal multiple access with cooperative relaying: a new wireless frontier for 5G spectrum sharing. IEEE Commun. Mag. **56**(4), 188–195 (2018)
7. Lin, Z., Liu, J., Ming, X., Gang, W., Liang, Y., Li, S.: Performance analysis and optimization in downlink NOMA systems with cooperative full-duplex relaying. IEEE J. Sel. Areas Commun. **35**(10), 2398–2412 (2017)
8. Shen, X.: Device-to-device communication in 5G cellular networks. IEEE Netw. **29**(2), 2–3 (2015)
9. Lee, J., Lee, J.H.: Performance analysis and resource allocation for cooperative D2D communication in cellular networks with multiple D2D pairs. IEEE Commun. Lett. **23**(5), 909–912 (2019)
10. Zhang, Z., Zheng, M., Ming, X., Ding, Z., Fan, P.: Full-duplex device-to-device-aided cooperative nonorthogonal multiple access. IEEE Trans. Veh. Technol. **66**(5), 4467–4471 (2017)
11. Kim, J., Lee, I., Lee, J.: Capacity scaling for D2D aided cooperative relaying systems using NOMA. IEEE Wirel. Commun. Lett. **7**(1), 42–45 (2018)
12. Huang, R., Wan, D., Ji, F., Hua, Q., Chen, F.: Performance analysis of NOMA-based cooperative networks with relay selection. China Commun. **17**(11), 111–119 (2020)
13. Qiao, Y., He, Y., Zhang, L., Yang, J., Zhou, L.: Performance analysis of multiple primary users CR-NOMA networks under partial relay selection. In: Proceedings of the 3rd International Conference on Wireless Communications and Applications (ICWCA), pp. 323–329. Hainan, China (2019)

Nonbinary LDPC Coded OTFS System Over Mobile Multipath Channels

Yiming Lu, Lin Zhou, Lin Wang, Sanya Liu, and Chen Chen

Abstract As communication technology advances, it becomes more difficult to keep up, low-density parity-check (LDPC) code has become one of the key technologies in channel coding with its superior error correction performance and efficient decoding algorithm. Among them, nonbinary LDPC codes perform better in high-order modulation and burst error channels, especially in short to medium code length conditions, and the tanner pattern of nonbinary LDPC codes is more sparse than that of binary LDPC codes and has a larger envelope length, which is more favorable to the optimization and design of LDPC codes in this form. In this paper, the basic principle and performance of the new waveform technology, for future 6G mobile communication, orthogonal time–frequency space (OTFS) modulation technology is addressed and examined in depth. In the high-band and high-speed mobile scenario based on the choice of nonbinary LDPC as the channel code. The results show that the OTFS modulation technique has better robustness, the lower peak-to-average ratio, and the potential to obtain full diversity gain on nonbinary LDPC channel coding compared with the conventional OFDM modulation system.

1 Introduction

In 1962, Dr. Gallager proposed the low-density parity-check (LDPC) codes [1], but its implementation was not possible due to the backward conditions at that time. Mackay and Neal discovered that before 1995, the capacity of the system with LDPC decoding algorithm in the case of longer code length is nearer to the Shannon limit than that of the Turbo code system. At present, the design, construction, performance analysis and application of the majority of LDPC codes are binary codes, while related studies

Y. Lu · L. Zhou (✉) · S. Liu · C. Chen
Xiamen Key Laboratory of Mobile Multimedia Communications, College of Information Science and Engineering, National Huaqiao University, Xiamen 361021, Fujian, China
e-mail: linzhou@hqu.edu.cn

L. Wang
China Mobile Communications Group Xiamen Co., Ltd. Xiamen, Fujian 361000, China

L. C. Jain et al. (eds.), *Advances in Wireless Communications and Applications*, Smart Innovation, Systems and Technologies 299,
https://doi.org/10.1007/978-981-19-2255-8_5

prove that nonbinary codes outperform binary codes in terms of performance under multi-channel conditions, nonbinary codes have much stronger burst error tolerance than binary codes. The nonbinary LDPC also has the following advantages: (1) it has the ability to eliminate small loops; (2) it can synthesize multiple burst errors into fewer multivariate symbol errors, thus improving the resistance to burst errors, which has great potential in 5G and even 6G systems [2]. In addition, combining the coding technology of nonbinary LDPC codes with multi-antenna systems can significantly increase the user capacity in the field of communication.

With regard to modulation waveform, 4G and 5G use the orthogonal frequency division multiplexing (OFDM) technology. In scenarios with a lot of movement, supported by B5G/6G, such as high-speed trains, vehicle to vehicle (V2V), unmanned aerial vehicle (UAV), satellite communications, etc., the high speed movement will generate large Doppler shifts, and the orthogonality between subcarriers of OFDM will be severely damaged, leading to a dramatic deterioration of performance. Recently, Hadani et al. proposed a brand-new waveform method, OTFS modulation, for high mobility scenarios [3]. Compared with the conventional OFDM modulation technique, OTFS employs a delay-Doppler domain signal representation that takes full advantage of the invariance, separability, as well as the orthogonality of the delay-Doppler domain channel's coupling to information symbols, and has the potential for full diversity gain and good robustness [4]. In this paper, we discuss the peak to average power ratio (PAPR), diversity gain, and coding gain of OTFS in conjunction with nonbinary LDPC channel coding for 6G high mobility scenarios and high frequency band communication scenarios, and discuss its advantageous potential and the issues that need further research and solution.

2 Nonbinary LDPC Codes

2.1 *Basic Concept*

GF(q) means a Galois domain with q members, with q deriving from prime numbers' powers. The check matrix $H = [h_{i,j}]_{0 \le i \le m, 0 \le j \le n}$, defined on the finite field GF(q) constitutes a regular LDPC code consisting of the following structural characteristics: (1) fixed column and row weights; (2) no identical non-zero values exist at the corresponding positions of every two rows (or two columns)[5]. If the row weight column weight is not unique, a non-regular LDPC code is described as the code word. The binary LDPC code's check matrix is made up of "0" and "1," but the nonbinary LDPC code's check matrix is made up of domain elements in a finite domain, for example, under the Galois domain GF(2^3), the domain elements include ($\alpha^{-\infty} = 0, \alpha^0, \alpha^1, \ldots \alpha^6$), where α denotes the native element under the domain element, and all domain elements can be represented by multiple powers of α. There are two main representations of nonbinary LDPC codes, check matrix and Tanner diagram. The check matrix's non-zero components are represented as label values

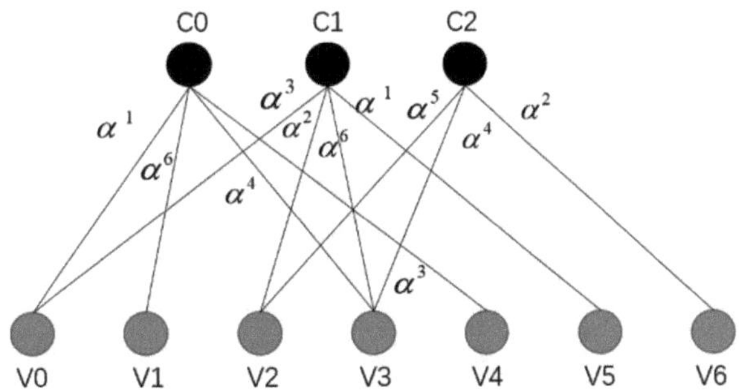

Fig. 1 Tanner graph of a nonbinary LDPC

on the connected edges in the Tanner diagram. The Hq of the nonbinary LDPC code defined in GF(2^3) is given in the following equation. The corresponding Tanner diagram is shown in Fig. 1.

$$H_q = \begin{bmatrix} \alpha^1 & \alpha^6 & 0 & \alpha^4 & \alpha^3 & 0 & 0 \\ \alpha^3 & 0 & \alpha^2 & \alpha^6 & 0 & \alpha^1 & 0 \\ 0 & 0 & \alpha^5 & \alpha^4 & 0 & 0 & \alpha^2 \end{bmatrix} \tag{1}$$

2.2 *Construction*

Nonbinary LDPC codes may be classified into two types of building methods, namely, randomized construction methods and structured construction methods. The progressive edge growth (PEG) algorithm is one of the commonly used randomized construction methods, which can gradually establish the link between check and variable nodes under known degree distribution, and is used to establish the Tanner graph of nonbinary LDPC codes with larger ring length. The Approximate Cycle EMD (ACE) algorithm is also one of the well-known computer-based random construction algorithms, which increases the influence of overlapping rings on the decoding code based on the consideration of rings, as a way to improve the flow of extra messages during decoding iterations. The above random construction method requires a large number of computer search operations, and constructed code words are irregular. Although the randomly constructed LDPC codes have good performance, the coding complexity is high [6].

3 New Waveform Technology—OTFS Modulation

For high mobility scenarios in 6G and high frequency band communication scenarios such as the millimeter wave, OTFS modulation technology can extend each information symbol in the time-delayed-Doppler domain to the entire time–frequency domain using the inverse symplectic finite fourier transform (ISFFT) compared to OFDM modulation technology. As a result, each transmitted symbol has a nearly constant channel gain with considerable resilience.

Figure 1 depicts the basic block diagram of the OTFS modulation approach. Assume a data burst packet has a total duration of NT seconds and the total bandwidth of MΔf Hz, with N being the number of OTFS symbols, M being the number of OTFS subcarriers, T being the symbol time interval, and Δf being the subcarrier frequency interval. The modulation symbols of length MN obtained from the message sequence u by constellation mapping are arranged into a two-dimensional matrix $x \in \mathbb{C}^{N \times M}$, which is the vector of information symbols on the DD plane, effective symbols at the k-th Doppler and l-th time-delayed grid point on the time-delayed-Doppler grid: $x[k, l] \in x, 0 \le k \le N-1, 0 \le l \le M-1$. First, the transmitter uses ISFFT to map the message symbols $X[n, m], 0 \le n \le N-1, 0 \le k \le M-1$:

$$X[n, m] = SFFT^{-1}(x[k, l]) = \frac{1}{\sqrt{NM}} \sum_{k=0}^{N-1} \sum_{l=0}^{M-1} x[k, l] e^{j2\pi(\frac{nk}{N} - \frac{ml}{M})} \tag{2}$$

The Heisenberg transformation may then be used to retrieve the time domain signal as follows:

$$s(t) = \sum_{n=0}^{N-1} \sum_{m=0}^{M-1} X[n, m] g_{tx}(t - nT_t) e^{j2\pi m \Delta f (t - nT_t)} \tag{3}$$

$g_{tx}(t)$ is the filter responsible for pulse shaping.

The linear time varying (LTV) channel [7] is represented for the DD plane class as:

$$h(\tau, v) = \sum_{i=1}^{P} h_i \delta(\tau - \tau_i) \delta(v - v_i) \tag{4}$$

where P denotes the number of pathways, and the ith path's path gain, time delay, and Doppler shift are indicated by h_i, τ_i and v_i. The ith path's time delay and Doppler shift may be represented as:

$$\begin{aligned} \tau_i &= \frac{l_{\tau_i}}{M\Delta f} \\ v_i &= \frac{k_{v_i} + \kappa_{v_i}}{NT} \end{aligned} \tag{5}$$

where l_{τ_i} and k_{v_i} denote the tapping parameters with relation to the time delay expansion and Doppler shift, respectively.

The above-mentioned channel receives the time plane signal as:

$$r(t) = \iint h(\tau, v)s(t-\tau)e^{j2\pi v(t-\tau)}d\tau dv + \overline{w}(t) \tag{6}$$

Next, the Winger transform (Heisenberg inversion) yields the time–frequency domain signal as:

$$Y[n,m] = \int r(t)g^*_{rx}(t - nT_t)e^{-j2\pi m\Delta f(t-nT_t)}dt \tag{7}$$

The equivalent symplectic finite fourier transform (SFFT) may be used to get the received signal in the DD plane.:

$$y[k,l] = SFFT(Y[n,m]) = \frac{1}{\sqrt{NM}}\sum_{n=0}^{N-1}\sum_{m=0}^{M-1} Y[n,m]e^{-j2\pi(\frac{nk}{N}-\frac{ml}{M})} + w[k,l] \tag{8}$$

where w is the sampling point of the corresponding DD domain noise.

As seen in the graph, OTFS can also obtain full partial set gain in time and frequency while maintaining a lower PAPR, which is clearly better compared to the PAPR of OFDM [8], and the figure depicts a simple example of the PAPR comparison of OTFS with OFDM. Where the horizontal coordinate is the threshold value of the PAPR in dB form and the vertical coordinate Pr is the probability that either OFDM or OTFS is greater than the threshold value. The PAPR of OTFS is much lower than the PAPR of OFDM for the same.

4 The Simulation Result of Nonbinary LDPC Coded OTFS

The message vector received by the receiver is $y_n = (y_{n,0}, y_{n,1}, \ldots, y_{n,b-1})$, the vector length is b, and each vector corresponds to a transmitted code element as x_n, then:

$$L(x_n = a) = \log\frac{p(y_n/x_n = a)}{p(y_n/x_n = 0)} \tag{9}$$

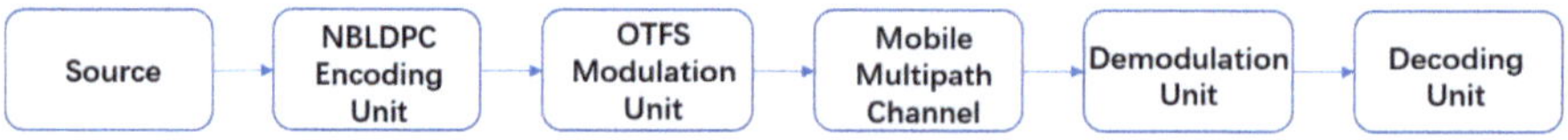

Fig. 2 Communication simulation system nonbinary LDPC decoding

The logarithmic domain Q sum product algorithm (QSPA) algorithm is as follows [9]: initialization first,

$$L(x_n) = \ln \frac{p(y_n/x_n = a)}{p(y_n/x_n = 0)} \tag{10}$$

$$L(q_{mn}^a) = L(v_n), L(r_{mn}^a) = 0 \tag{11}$$

Then the message update of the check node and the message update of the variable node are performed:

$$L(r_{mn}^a) = L(h_{m,n_m}^{-1} \sigma_{m,n_{m,l}} + h_{m,n_m}^{-1} \rho_{m,n_{m,l}}) \tag{12}$$

$$L(q_{mn}^a) = L(x_n) + \sum_{m' \in M(n) \backslash m} L(r_{mn}^a) \tag{13}$$

The final judgement, similar to the standard SPA decoding, is that after the sign judgement, the decoding of the all-zero vector of the accompanying equation is completed, otherwise it returns to the next iteration until the accompanying equation is satisfied or the preset iteration maximum is reached, and the decoding is completed, reaching the desired result [10] (Fig. 2).

We use nonbinary LDPC code in GF(16) domain for simulation in channel coding, which has a 512-byte code length and a 1/2 coding rate, where the number of symbols of OTFS N = 8, the number of subcarriers M = 16, and the moving speed is 250 km/h. The graph compares the performance of OTFS modulation with OFDM modulation under 16QAM and 256QAM, and the OTFS condition can also be found in the diagram. The simulation results show that with 16QAM modulation, the OTFS system achieves a gain of 1.8 dB over the OFDM system under multipath fading channel, and similar simulation results are obtained when modulated with 256QAM (Figs. 3 and 4).

5 Conclusion

Under mobile multipath channel, OTFS became a waveform technology with development potential in 6G mobile communication by virtue of its excellent full diversity potential, low PAPR, and good robustness. In this paper, this modulation technique is combined with nonbinary LDPC channel coding. On the one hand, the construction

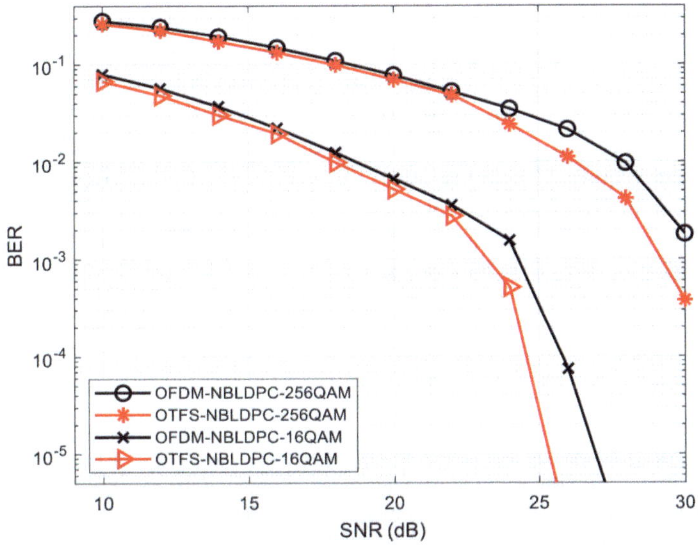

Fig. 3 BER performance under different modulation method

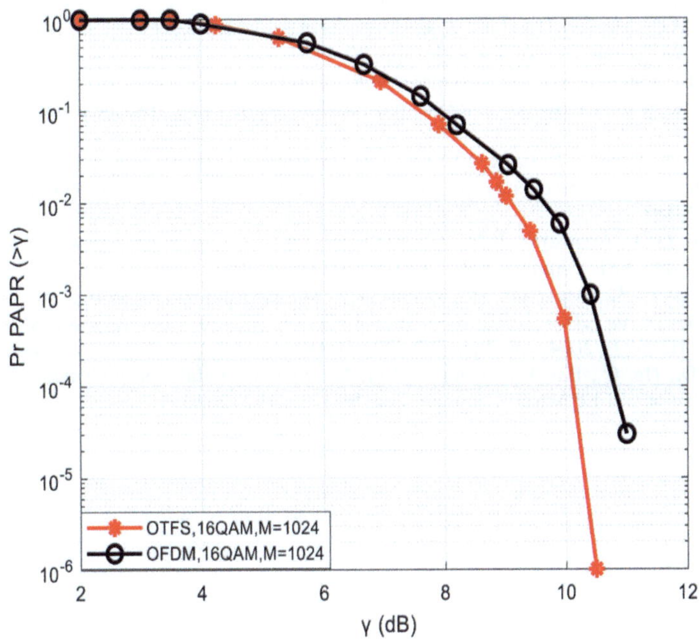

Fig. 4 PAPR comparison of OTFS and OFDM

compilation code process of nonbinary LDPC and OTFS principle are analyzed. On the other hand, the simulation systems of both are established. Nonbinary LDPC codes' performance with various modulation schemes in multipath fading channels is simulated using the QSPA decoding algorithm, and the system BER graph and the PAPR comparison graph of OTFS are obtained. Since the OTFS system has the important feature of being relatively insensitive to time changes, high-motion sceneries and multipath fading scenes are good candidates for the OTFS technology. In the future, the system can be further optimized in terms of channel estimation and the application of different channel coding.

References

1. Gallager, R.: Low-density parity-check codes. IRE Trans. Inf. Theor. **8**(1), 21–28 (1962)
2. You, X., Wang, C X., Huang, J.: Towards 6G wireless communication network. Vision, enabling technologies, and new paradigm shifts. Sci. China Inf. Sci. **64**(1), 5–78 (2021)
3. Hadani, R., Rakib S., Tsatsanis, M.: Orthogonal Time Frequency Space Modulation. arXiv: 1808.00519v1 (2018)
4. Tiwari, S., Das, S.S., Rangamgari, V.: Low complexity LMMSE receiver for OTFS. IEEE Commun. Lett. **23**(12), 2205–2209 (2019)
5. Davey, M.C., Mackay, D.J.C.: Low density parity check codes over GF(q). In: 1998 Information Theory Workshop (Cat.No.98EX131), pp. 70–71 (1998)
6. Rong, B., Jiang, T., Li, X., Soleymani, M.R.: Combine LDPC codes over GF(q) with q-ary modulations for bandwidth efficient transmission. IEEE Trans. Broadcasting **54**(1), 78–84 (2008)
7. Raviteja, P., Phan, K.T., Hong, Y., Viterbo, E.: Interference cancellation and iterative detection for orthogonal time frequency space modulation. IEEE Trans. Wirel. Commun. **17**(10), 6501–6515 (2018)
8. Surabhi, G.D., Augustine, R.M., Chockalingam, A.: Peak-to-average power ratio of OTFS modulation. IEEE Commun. Lett. **23**(6), 999–1002 (2019)
9. Wymeersch, H., Steendam, H., Moeneclaey, M.: Log-domain decoding of LDPC codes over GF(q). In: 2004 IEEE International Conference on Communications (IEEE Cat. No.04CH37577), 2004, vol. 2 pp. 772–776 (2004)
10. Feng, D., He, Q., Bai, B., Zheng, J., Liu, M.: Spatial modulation with multi-dimensional constellations. IEEE Wirel. Commun. Lett. **9**(1), 99–102 (2020)

Exploration of Environmental Response Techniques for Land Reclamation in Caofeidian

Dong Du, Hongwei Liu, Xiaoming Sun, Futian Liu, Hui Kang, and Xinyu Cao

Abstract The rational construction and planning of coastal cities has always been an important direction for urban development. The geological environment and sea level changes of the coastline are closely related to the safety of people's property. The interaction of reclamation projects and the changes brought about by ocean hydrodynamic conditions have a significant impact on the coastal zone. The geological environment has had a serious impact. Based on the environmental geological survey and evaluation results of the coastal zone in the Caofeidian New Area in recent years, this paper focuses on the land restoration of coastal cities by analyzing changes in coastlines and lake beaches, changes in tidal currents, coastal erosion and alluvial layers, storm surges and sea level changes. The interaction and impact of reclamation and the geological environment, and corresponding prevention and control suggestions for major issues such as storm surge and sea level rise, and scientific guidance for reasonable planning and disaster warning of coastal cities.

1 Introduction

Caofeidian has become an artificial land connected to land island from the original 4 km^2 island. An ecological, new and modern industrial park and coastal city have been born on the coast of Bohai. The reclamation area of the Caofeidian industrial area has reached 247 km^2 (see Fig. 1); The planning area of Caofeidian ecocity is 150 km^2, population size will reach 1 million people, of which the first phase of the

D. Du (✉) · H. Liu · X. Sun · F. Liu · H. Kang
Tianjin Center of Geological Survey, CGS, Tianjin 300170, China
e-mail: yndd24@163.com

H. Liu
e-mail: 61346723@qq.com

X. Cao
Laboratory Cultivation Base of Environment Process and Digital Simulation, Capital Normal University, Beijing 100048, China

L. C. Jain et al. (eds.), *Advances in Wireless Communications and Applications*,
Smart Innovation, Systems and Technologies 299,
https://doi.org/10.1007/978-981-19-2255-8_6

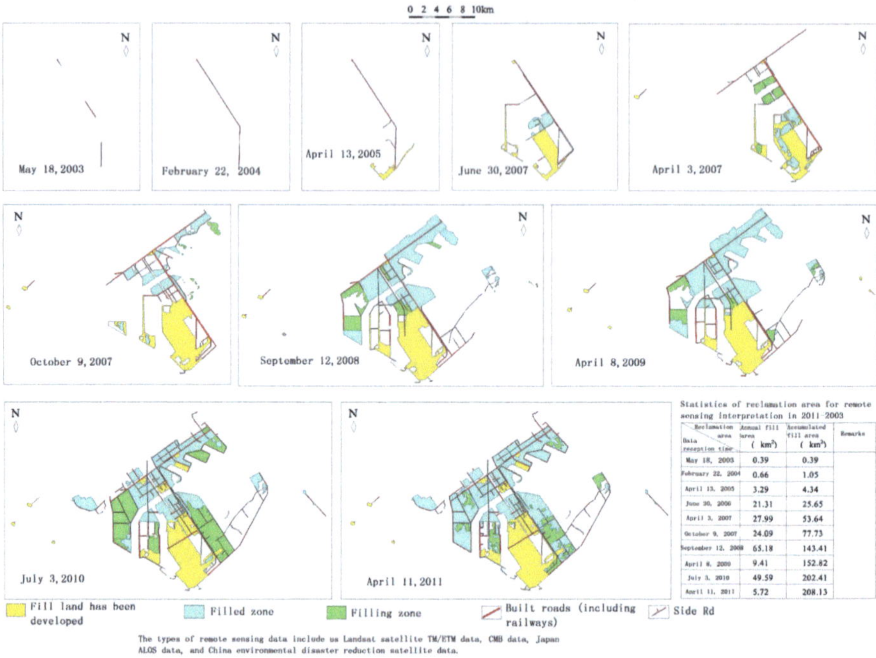

Fig. 1 Remote sensing interpretation map of land reclamation from sea in Caofeidian Industrial Zone

project is a part of the sea reclamation, and gradually to the sea by land construction (see Fig. 2).

2 The Changes and Countermeasures of Land Reclamation Caused the Coastline and Intertidal Zone

Sea reclamation interaction and ocean hydrodynamic conditions changed the composition gradient and the material type, length of coastline and intertidal zone, has a serious impact on the geological environment in the coastal zone. The coast of Caofeidian new area is composed of muddy coast, sandy mud coast, sandy coast, sandy coast and artificial shore dike with different revetment types (see Fig. 3).

The coast protection embankment types of Caofeidian eco city are mainly natural earth dams before large-scale construction, including natural earth dams, gravel and natural dams, gravel and geotextiles. There are few surface buildings, mainly aquaculture area, the vegetation is poor, and the sediment types from west to East are mainly argillaceous and Sandy. Basically, there is no protection for any dams. The local part is combined with gravel and geotextile to protect slope (see Fig. 4), the

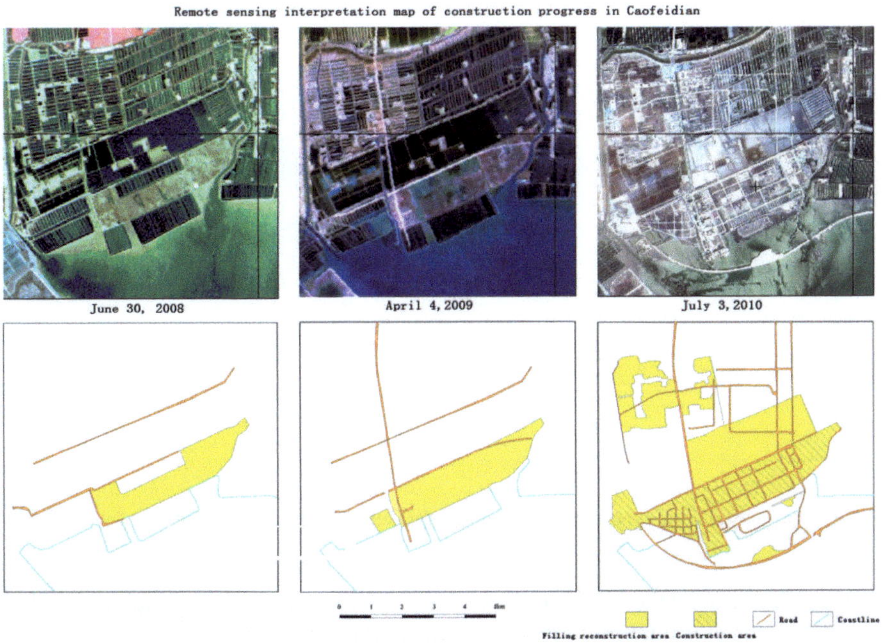

Fig. 2 Remote sensing interpretation map of construction progress in Caofeidian International Eco City

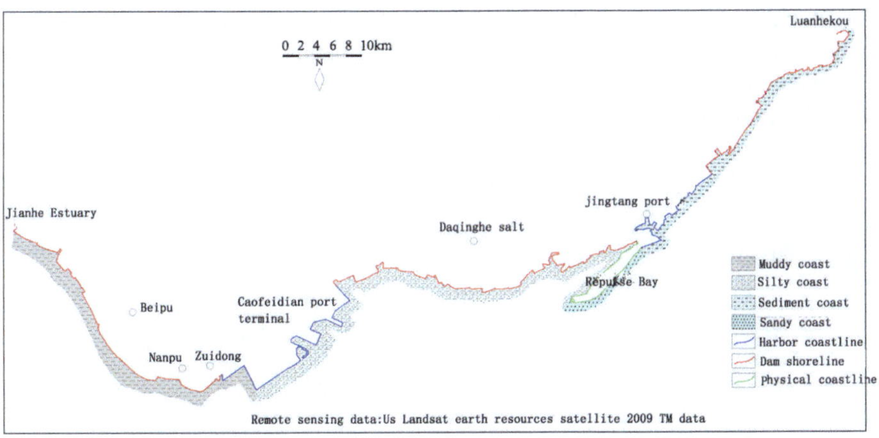

Fig. 3 The coastline type map of Caofeidian in 2009

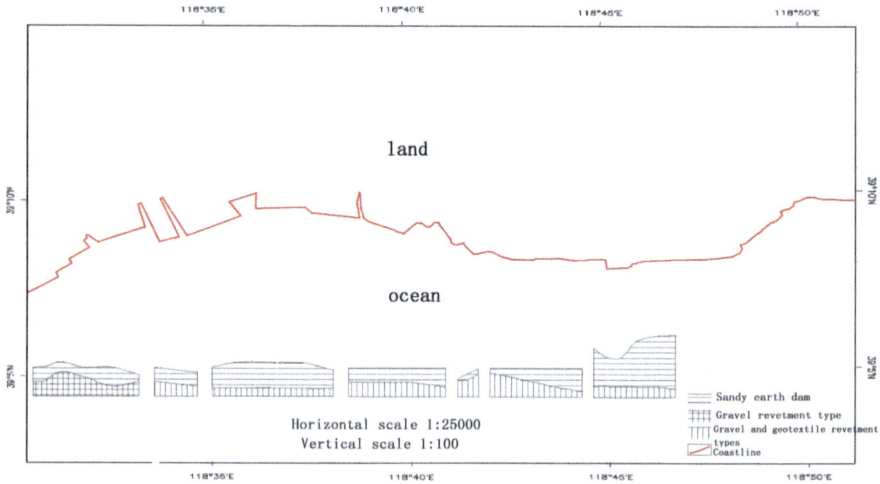

Fig. 4 The Coastal berm type zoning map in Caofeidian International Eco City

average elevation on the top of the coastline is 3.64 m (see Fig. 5), which is very low in resisting the storm hydrodynamic and wave erosion.

According to the investigation, in the past thirty years, because of Yantian, aquaculture and continue to expand the scale of Caofeidian coastline is advancing to the sea, shoreline length increased 60 km, the land area increased 304 km^2, intertidal current width of 2.5–6.6 km, but showed a trend of gradually narrowing the width of the beach face. According to Yang Yanxiong et al. Comparison of Caofeidian reclamation engineering 1983–1996 thirteen years ago, the average bank accumulation rate is 3.2 cm/a, the slope erosion rate under water 7.2 cm/a. According to the 2008–2010 beach erosion dynamic monitoring data show that after the implementation of large-scale Caofeidian sea reclamation project, to the East are to scour from the

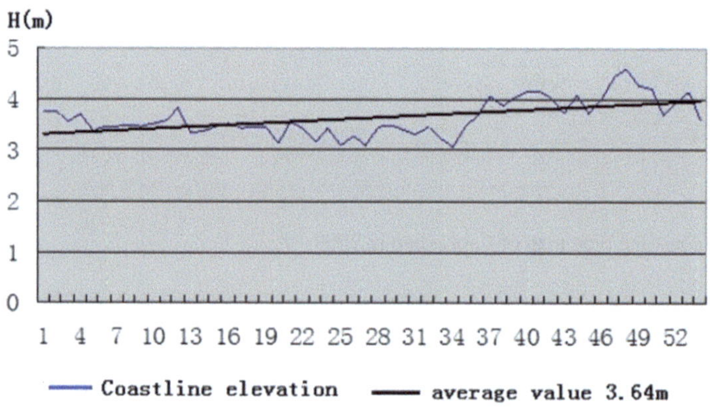

Fig. 5 The coastline top elevation map in Caofeidian International Eco City

intertidal zone, the erosion rate is generally 5–20 cm/a (comprehensive geological survey report of Jingtang Port, Yang Yanxiong et al. (2003)). In this regard, according to Caofeidian marine hydrodynamic conditions and coastal beach erosion, sediment dynamics, scientific layout of land reclamation and coastal engineering, reduce the disturbance to the coast and beach surface, avoid its sharp change.

3 Reclamation of Tidal Current Field and Seabed Caused by Impact on the Changes and Countermeasures

Under the natural condition that the Caofeidian industrial area has not been developed on a large scale, the tide of the Caofeidian sea is a reciprocating flow: In Caofeidian the west to the East, the ebb. Caofeidian to form Cape south into the Gulf of Bohai, the main direction is not the same, but the law is obvious, the mainstream direction basically parallel to the Isobaths, namely Caofeidian the west to west north to East biased towards the south, the ebb tide in the West Austin, Caofeidian slightly southward, ebb tide to the Northeast (see Fig. 6).

According to the studies of Kuang Cuiping et al., Caofeidian industrial zone through Island Road (dam) after the completion of the Caofeidian sea area in deep water area of tidal field did not change significantly, just cut off the channel embankment East and West groove, beach water is divided into two channel embankment East and west part, diffuse Beach flow cannot converge at high tide beach ridge, and converted to current along Breakwater (see Fig. 7).

But the flow field near the reclamation area has changed. The tidal current along the Caofeidian East Sea has been leveed towards the Caofeidian sea head. After the water bypass the Caofeidian head, it is affected by the impact of the pick up flow, and the water flows along the deep trough of the front edge of the Diandian to the northwest. Influence of breakwater, velocity into the excavated area obviously

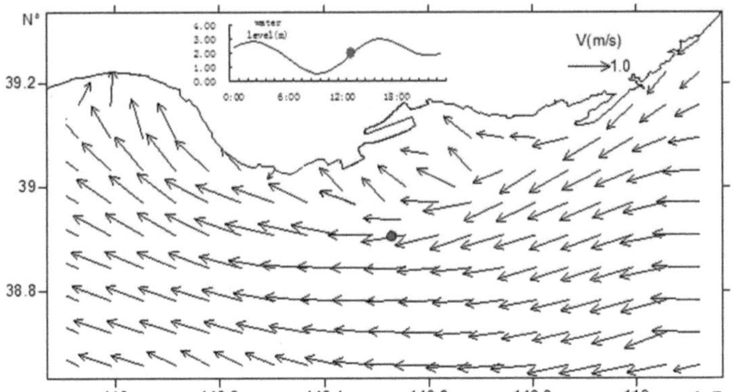

Fig. 6 Rapid flow field map before the construction of the Island Road (embankment) in Caofeidian

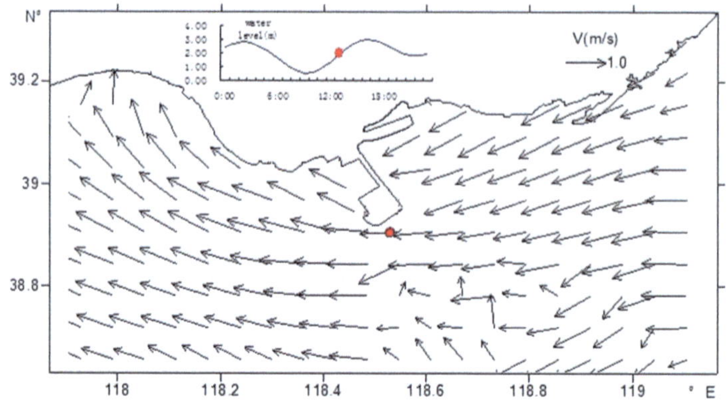

Fig. 7 Rapid flow field map after the construction of the Island Road (embankment) in Caofeidian

slowed down, a large amount of water around the east engineering reclamation area into shoal until through Island Road (dam); the ebb, West shallow water flows to the excavated harbor basin, most still from the west is not included in the reclamation areas deep trough along the deep channel from the west to the northwest, flowing into the heavy head, the head heavy deflecting effect to the northeast to flow. Thus, the rise and fall trend is significantly affected by the deflecting effect has heavy head reclamation headland, so both sides of velocity significantly affected (see Fig. 8). Yin Yanhong [1] suggested that through Island Highway through the shoal tidal channel should be built bridge, to preserve the Caofeidian shoal tidal channel and the main trend of marine protection system. The reclamation of soil in the reclamation area should not be carried out near the barrier island of Caofeidian and its deep trough in the front, protecting the bottom of these key parts from damage and maintaining the function of the port.

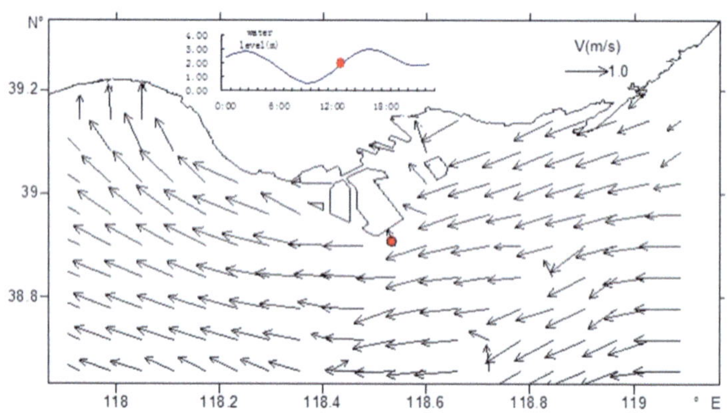

Fig. 8 Rapid flow field map of the current condition of the enclosed Engineering in Caofeidian

Caofeidian port site is the only one in Bohai does not require excavation of waterways and basins can build 300 thousand ton berths large natural harbour, with excellent depth, large and stable, berth trough to anchor well in the Gulf of Bohai incomparable natural advantages. Due to land reclamation in Caofeidian Industrial Zone and Island Road (DI) construction, exacerbated by the heavy head Cape effect, caused by tidal flow field changes Heavy head deep groove bypass ahead of time about 20 min, the average velocity of flood and ebb tide increased, an increase of 1.8–31.4%, the increase of 4.5–30.2%, resulting in 0.1–0.5 m ranging from scouring Caofeidian front range of 3, 5 km and slight erosion until The west side of the plateau has a maximum erosion of more than 0.5 m. There is about 0.2 m deposition in the 3 km area of the western front of the Caofeidian project, while 5–6 km is slightly eroded outside, and the local scour is relatively large. The maximum siltation of about 0.3 m near the port gate of a port has a maximum of about 0.7 m deposition near the gate of the two ports. By the wave action, Western and Eastern Beach Caofeidian respectively Shatuo is greater than that of 1.2 and 1.4 m in Western beach erosion; erosion results in front of deep groove 0.1–0.4 m ranging from deposition, while the eastern Shatuo led to the erosion of laolonggou deep trough 0.12–0.53 m deposition (see Fig. 9).

Impact of the construction of Caofeidian port engineering of sea wave field is very obvious, after the completion of the project, harbor pool, tidal channel in the wave field changed significantly, the wave height and period significantly reduced, controlled by the incident wave to change to engineering and trend, and for the wave characteristics of sea engineering outside the area has little influence on [2].

According to the dynamic monitoring of 2008–2010 four profiles, Caofeidian West to a lesser extent; continuous siltation, heavy head east main scour, near laolonggou sea monitoring points to micro in main.

According to the Caofeidian deep trough to the north and South ADCP walking and sailing velocity measurement, vertical upward velocity decreases with the

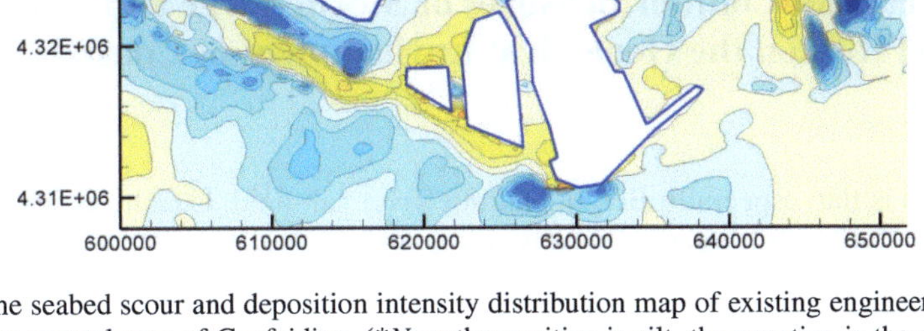

Fig. 9 The seabed scour and deposition intensity distribution map of existing engineering conditions in the coastal area of Caofeidian. (**Note* the positive is silt, the negative is the scour, unit: m)

increase of water depth, the strongest Caofeidian deep groove flow profile, on the one hand it is the main power to maintain the Caofeidian deep groove, on the other hand, is also one of the factors causing the Caofeidian Industrial Zone water slope instability. According to the data analysis, the existing engineering construction although the cause of marine hydrodynamic conditions changed, Caofeidian slightly erosion, but the slope under heavy water remained relatively stable, yet obvious erosion.

Compared with the survey data before 1996, the surface sediments of the seabed are in a certain degree of coarsening. From 2008 to 2010, affected by the reclamation and marine hydrodynamic conditions, suspended sediment content in Caofeidian sea area has been reduced year by year, which in 2009 decreased by 15% compared with 2008, 2010 is about 39% less than in 2008.

The change of tidal current caused Caofeidian coastal erosion and port and waterway rush to destroy coastal protection works and other coastal facilities such as seawall. The loss of coastal land not only threatened the safety of Caofeidian industrial area and International Eco-City, but also increased the deposition of ports and waterways. Superimposed with other disasters such as storm surge and land subsidence, the disaster aggravated further, and reduced the stability of coastal engineering foundation, which seriously affected the safety of the geological environment and the sustainable development of economy and society in coastal cities [3, 4]. The change of scouring and silting of port and waterway in the area has led to deep groove topography changes and underwater slope stability, which is a significant geological environmental problem in the development of Caofeidian new area. With the continuous development of Caofeidian port construction, further progress of effects of reclamation and port engineering construction of marine power, suspended sediment distribution and seabed evolution will. Therefore, should strengthen the Caofeidian marine hydrodynamic environment, Caofeidian deep trough siltation (especially tidal river open, the tidal current field can restore to the original state, heavy head deep channel trend will continue to increase), underwater slope stability monitoring and research, closely tracking the interaction and land reclamation effect of engineering and layout and flow field, it will affect the safety of port engineering, to further optimize the land reclamation to provide the scientific basis for the safe operation of the project scale and layout as well as the Caofeidian Industrial zone.

4 Effects and Countermeasures of Storm Surge and Sea Level Rise on the Geological Environment of Coastal Zone

The region is flat, especially Beipu, Daqing River salt field, Luanhe estuary, coastal salt shoals, Qilihai fishing port, because there is no seawall to prevent or levee elevation is low, is the key areas vulnerable to storm surge disaster damage. As part of the flood diversion areas of Luan River, no standard seawall to prevent, by the influence of economic development, the mouth also has large tracts of farmland and ponds, vulnerable to storm surge; Seven the Caspian Sea fishing port, in 2003 during the "1011" storm surge, Sheung Shui pier near the houses flooded water about 1 m; The two part of the Luanhe River Estuary, because there is no effective construction of the seawall is the main region of the current storm surge inundation. There have been many storm tide disasters in the history area which have a great influence on the geological environment of the coastal zone. According to historical records, in 1895 and 1900, there were two strong storm surges attacking Caofeidian. The temples and shops on the island were all washed into the sea, and the migration of the islanders gradually cooled down. According to statistics from the past twenty years, 0.9 waves of wave height greater than or equal to 6 m occur every year in Bohai, including 9 waves of cold waves, 12 cyclones and 2 waves of typhoons [5].

The average elevation of Caofeidian coastline is about 4 m, the lowest elevation is located in the coastal shoreline of Beipu Jianchang, Daqing River, the average height less than 3 m, which is lower than the local coastal storm surge warning level value 3.8 m and value 3.65 m the highest tide level (see Fig. 10). Due to the influence of ground subsidence and other factors, the elevation of the coastline in 2009 is lower than that of the 2008 shoreline elevation, but the decline of most of the banks is smaller. The coastline elevation of the Caofeidian industrial area is about 3.5 m, and the average height of the breakwater is about 4.75 m. Most of the coastline heights are lower than the local storm surge warning level. However, the height of the breakwater dike is mostly higher than the highest tide level and the Centennial tide value is 4.48 m.

According to this study of Yu Daoyong, Zhang Yecheng, and others, in the past 1980–2008 years, the coastal sea level of Caofeidian has fluctuated upward trend, which has increased by 118 mm on average for 28 years, and the annual rate of long-term change is 4.1 mm/a (see Fig. 14). In order to predict the future trend of sea level change, the regression equations (Table 1 and Figs. 11, 12, 13) are established by using the measured data of Caofeidian in the past two years and the synchronous

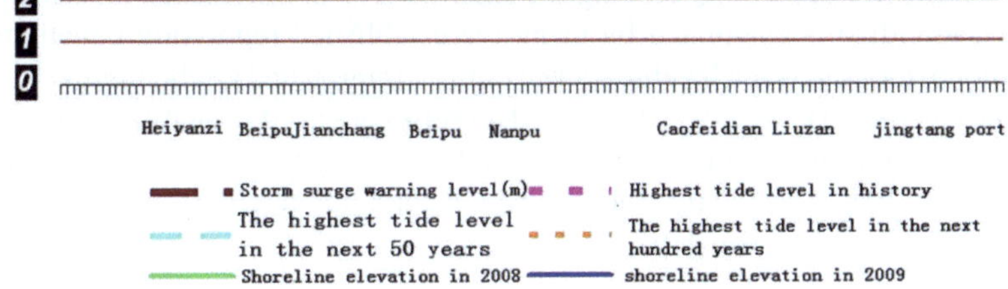

Fig. 10 The elevation curve variation chart of Coastline in Caofeidian

Table 1 Correlation coefficient of regression equation

Number of samples	Category	Correlation coefficient	Significant level	
			0.05	0.01
33	Monthly mean sea level between Caofeidian and Tanggu ϒ1	0.997	0.344	0.443
33	Monthly average maximum sea level between Caofeidian and Tanggu ϒ2	0.954	0.344	0.443
33	Monthly mean sea level between Caofeidian and Qinhuangdao ϒ3	0.993	0.344	0.443
33	Monthly average maximum sea level from Caofeidian to Qinhuangdao ϒ4	0.464	0.344	0.443

ϒ1–ϒ4 are significantly correlated, but ϒ4 is not good

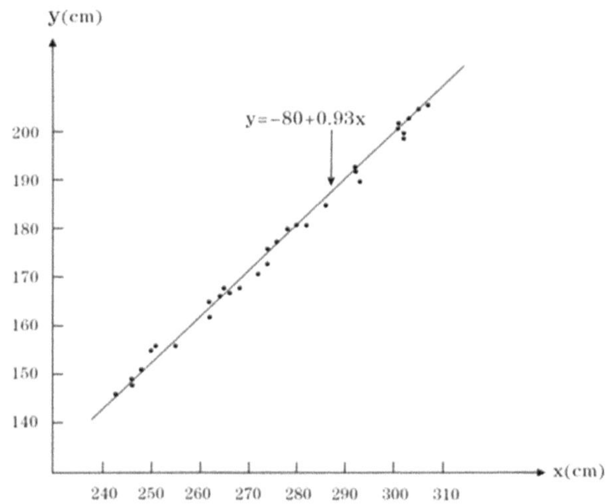

Fig. 11 Regression line of monthly average sea level between Caofeidian and Tanggu

data of nearby Tanggu and Qinhuangdao stations. The historical sea level sequence data of Caofeidian is calculated by using the established regression equations and the historical measured sea level sequence data of Tanggu and Qinhuangdao stations. Then, the calculated Caofeidian sea level data series are used to analyze its long-term change trend, amplitude, annual rate and multi-year high sea level, and predict the future change. It is expected that the sea level of Caofeidian will rise by 157–187 mm in the next 30 years than in 2010.

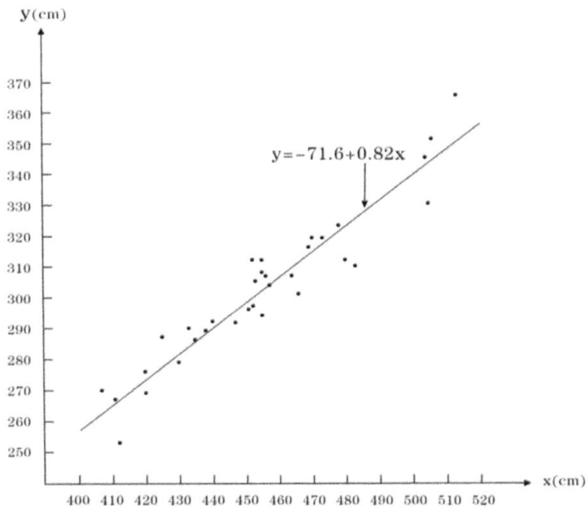

Fig. 12 Regression line of monthly average maximum sea level between Caofeidian and Tanggu

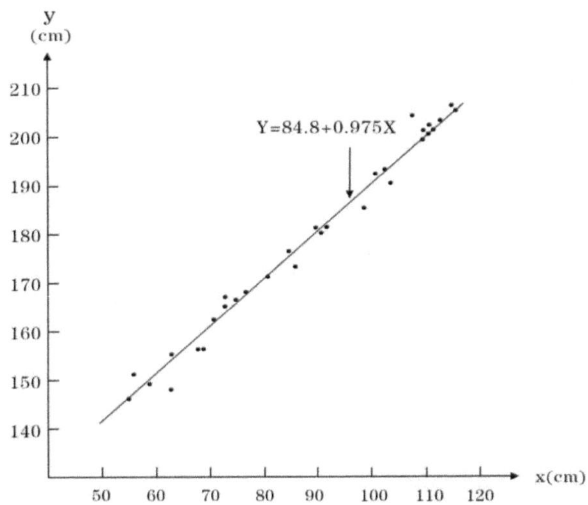

Fig. 13 Regression line of monthly mean sea level from Caofeidian to Qinhuangdao

Monthly mean sea level between Caofeidian and Tanggu:	$y = -80 + 0.93x$	(1)
Monthly average maximum sea level between Caofeidian and Tanggu:	$y = -71.6 + 0.82x$	(2)

(continued)

(continued)

Monthly mean sea level between Caofeidian and Qinhuangdao:	$y = 84.8 + 0.975x$	(3)
Monthly average maximum sea level from Caofeidian to Qinhuangdao:	$y = 134 + 0.1457x_1 + 0.0367x_2$	(4)

Thus, the ability to resist storm surge in some sections of the Caofeidian new area is relatively weak in the context of the continuous rising sea level. Caofeidian Tong Dao Road (dike), Shougang Group, petrochemical platform, oil and Ore Wharf and other important projects are the main facilities for the storm surge. From the Nanpu Economic Development Zone, Nanpu oil field, Caofeidian Industrial Zone, Caofeidian International Eco-City, to shijiutuo Island Tourism Zone, development zone, seven Caspian Sea tourist area and natural protection zones are needed Focus on prevention and protection. Therefore, when designing the coastal dyke, we should consider not only the beach type and the coastline change characteristics, but also the bank's functions, strength and top elevation to prevent extreme marine disasters, and optimize the layout and engineering design of the embankment. We should fully consider the ground elevation loss problem caused by ground subsidence, sea-level rise and other factors, when design the moisture-proof crest elevation (Fig. 14).

According to the factors such as ground elevation, 100 years' tide level, land subsidence rate and sea level rising rate, it is suggested that the top elevation design of the embankment of Caofeidian new area should be 6.79 m at one hundred years (the national 85 elevations). In order to avoid and mitigate the losses caused by storm surge disasters and safeguard the safety of people's lives and property, there are three ways to evacuate the storm surge: In the first case, a proposal for evacuation and transfer of high safety level for ground elevation control is proposed; The second is the suggestion of evacuation and transfer in the risk level area of the temperate storm flood; The third case is for the evacuation and transfer of typhoon storm flood risk level area, as shown in Fig. 15.

5 Conclusion

This paper analyzes the mutual influence of Caofeidian reclamation and coastal geological environment from the change of the coast and beach, tidal current field and seabed erosion, storm surges and rising sea levels. The reclamation of the new city has caused the change of sedimentary environment and hydrodynamic conditions of the ocean, changing the coast, beach, seabed erosion and siltation status. Storm surge and sea-level rise have intensified the impact on the geological environment of the coastal zone, which is a major issue to be paid attention to in the construction of the new area of Caofeidian. Suggestion:

(1) To study the degree of land reclamation and interaction influence engineering and flow field development, strengthen erosion dynamic monitoring of heavy

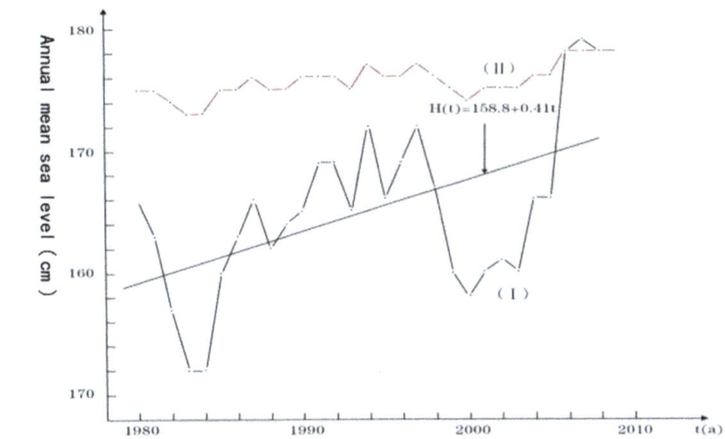

Fig. 14 The annual mean sea level change Diagram of tide gauge station in Caofeidian

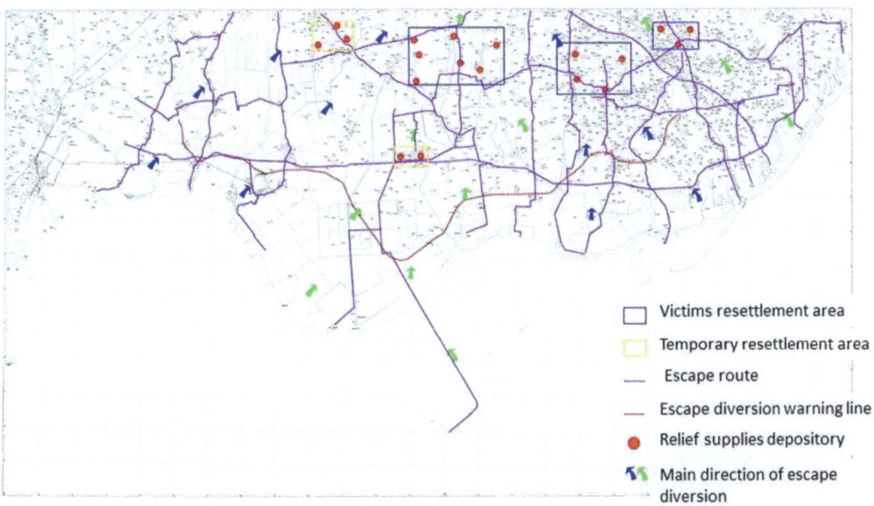

Fig. 15 The evacuation transfer sketch map of suffered storm surge disaster in Caofeidian

head submarine deep groove in Caofeidian sea area, to provide a scientific basis for further optimize the land reclamation engineering scale and layout.

(2) We should strengthen the dynamic monitoring and research of the underwater slope stability in key coastal areas such as the Caofeidian industrial area and International Eco city by using a variety of technical means and methods.

Fund Project

Geological survey project of the China Geological Survey (20,190,251); Special funding project of national nonprofit industry research (201,011,019); Monitoring and evaluation of resources and environment carrying capacity of Beijing Tianjin Hebei coordinated development zone and Xiong'an New Area (DD20221727).

References

1. Yin, Y.H.: Thinking about the large area reclamation of the Caofeidian shoal in Tangshan, Hebei. Mar. Geol. Dyn. 3 phase of 2007 (2007)
2. Gou, H.L., Liu, S.G., Kuang, C.P.: The effect of Caofeidian project on the Sea wave field. J. North China Water Conservancy Hydroelectric Inst. 5 phase of 2009 (2009)
3. Chen, M. X.: Geological environment characteristics and geological environment system in coastal areas—Also on "human land system". J. Geol. Disasters Prev. China 11(Suppl.), 81–86 (1998)
4. Gao, M.S., Zhu, Y.F.: Coastal Environmental hydrogeological problems and prevention in China. Mar. Geol. Dyn. **22**(5), 8–10(2006)
5. National marine forecasting center.: Special report on the study of storm surge disaster regionalization and forecasting and early warning in Caofeidian area (2011)

Spatially Coupled LDPC Codes Based Joint Source-Channel Coding

Lian Qiufang, Zhou Lin, Chen Qiwang, Chen Chen, and Wang Xi

Abstract Source coding and channel coding are the hot research parts in information theory. The source channel separation system cannot take full advantage of the correlation between source and channel. The performance of the joint source-channel coding (JSCC) system can be improved by using the redundant information of the source and this system is more suitable for future continuous error-free communication needs. In recent years, many applications require error-free transmission with low latency. Due to the special structure of parity check matrices (PCM), spatially coupled low-density parity-check (SC-LDPC) codes can adopt the sliding window decoding (SWD) algorithm to have the characteristics of low delay and ensure continuous error-free transmission in streaming media applications. Simulation results show that both spreading factor and window size affect decoding performance. This paper reviews the design of the sliding-window decoding algorithm in additive white Gaussian noise (AWGN) channel based on SC-LDPC codes and prospects future work.

1 Introduction

In the traditional communication system, the optimization of source encoder and channel encoder are two independent processes. And the traditional communication system is based on the principle of separation and optimization design. To satisfy the source-channel separation theory proposed by Shannon, two conditions need to be met: (1) the length of codeword sequence is infinite; (2) The whole system does not require the complexity of source and channel coding and decoding algorithms. However, these two conditions are virtually impossible to achieve. Borkenhagen and Saywood et al. proposed that redundant information existing in source data can be

L. Qiufang · Z. Lin (✉) · C. Qiwang · C. Chen
Xiamen Key Laboratory of Mobile Multimedia Communications, , College of Information Science and Engineering, National Huaqiao University, Xiamen 361021, Fujian, China
e-mail: linzhou@hqu.edu.cn

W. Xi
China Mobile Communications Group Fujian Co., Ltd. Fuzhou, Fujian 350003, China

L. C. Jain et al. (eds.), *Advances in Wireless Communications and Applications*, Smart Innovation, Systems and Technologies 299,
https://doi.org/10.1007/978-981-19-2255-8_7

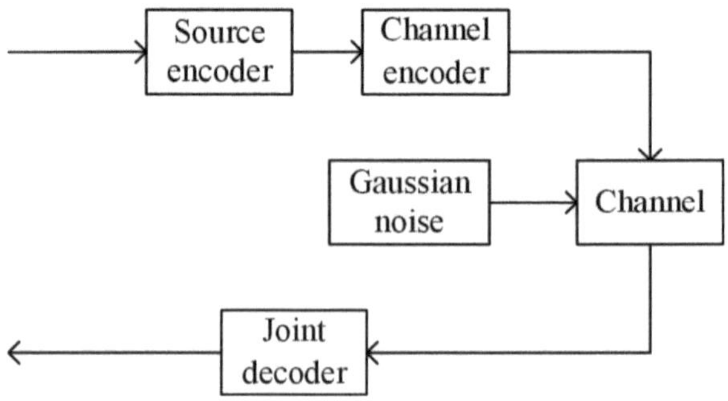

Fig. 1 The process diagram of the joint source channel coding system

used to optimize channel encoders [1]. Pettijohn et al. used the structure of two continuous decoders for the first time, and the practice proved that this structure has strong data recovery ability [2].

There are three source channel joint modes: (1) source and channel in the encoder union; (2) source and channel are combined in the decoder; (3) source and channel are combined in the encoder and decoder. The second combination method is used in this paper, and the block diagram of the whole system is shown as follows (Fig. 1).

Spatially coupled LDPC codes are the convolution codes of low-density parity-check codes [3]. SC-LDPC codes can be traced back to the convolutional LDPC codes proposed by Felstrom and Zigangirov in 1998 [4]. SC-LDPC codes have threshold saturation characteristics, and the decoding threshold under the belief propagation (BP) decoding algorithm can reach the maximum a posterior probability (MAP) of corresponding regular LDPC codes. Thus, Lentmaier proved that SC-LDPC codes can theoretically reach the Shannon limit [5]. To reduce the average decoding complexity of sliding window decoding, a predictive adaptive based on bit error rate (BER) is proposed window shift method [6]. Paper [7] proposed a method of introducing supervision to improve the performance of sliding window decoding.

The characteristic of convolutional codes is that the coded output bits of each code group are related both to the information bits of the group and but also to the information bits of other groups at the previous moment. SC-LDPC codes as convolution codes also have this characteristic. In terms of coding, SC-LDPC code can be encoded by syndrome former realization or partial syndrome former realization [8]. The second encoding method is used in this paper.

Theoretically, the performance of SC-LDPC codes will improve with the increase of coupling length. However, the longer the coupling length is, the greater the decoding delay and complexity of SC-LDPC codes will be. Iyengar et al. proposed that SC-LDPC codes could be decoded with a sliding window decoder, thus reducing the decoding complexity and delay [9].

2 SC-LDPC Codes

SC-LDPC codes are obtained by copying, edge spreading, and coupling regular LDPC codes. Take LDPC codes with degree distribution (3,6) as an example, as shown in Fig. 2, which shows how SC-LDPC codes are obtained by (3,6) LDPC codes. Figure 2 shows the protograph of (3,6) LDPC codes. In Fig. 2, a circle represents M variable nodes and a square represents M check nodes, where M is called spreading factor.

Firstly, L copies of the protograph unit as shown in Fig. (a) were made, and the position labelled t (0, ..., $L - 1$), L is called the coupling length. Then Fig. (b) shows the process of edge spreading. Finally, additional m_s check nodes are added to terminate edge spreading, where m_s is called coupling width.

In JSCC system, the coupling lengths of source SC-LDPC codes and channel SC-LDPC codes are represented by L^{sc} and L^{cc} respectively, where $L^{sc} + 2\ ms = L^{cc}$. In addition, the coupling widths of source SCLDPC code and channel SCLDPC code are equal and both are represented by m_s.

2.1 Encoding SC-LDPC Codes in JSCC

There are two LDPC codes in the JSCC system, as shown in Fig. 3. The PCM of source LDPC codes is $\mathbf{H}^{sc}$, the input sequence of the source encoder is **s**, and the output sequence after encoding is **u**. The second LDPC codes are channel LDPC codes, the PCM is $\mathbf{H}^{cc}$ [10].

The input sequence of source encoder is $\mathbf{s}_{[0,L-1]} = [\mathbf{s}_0, \ldots, \mathbf{s}_{L-1}]$ and probability of 1 in the source is 0.02.

The output sequence of source encoder is $\mathbf{u}_{[0,L-1]} = \mathbf{s}_{[0,L-1]}\ (\mathbf{H}^{sc})^{\mathrm{T}}$, $\mathbf{u}_i = (\mathrm{u}_i^{(1)}, \ldots, \mathrm{u}_i^{(M)})$. This is the process of source encoding.

The input and output of the channel encoder are $\mathbf{u}_{[0,L\text{-}1]}$ and $\mathbf{v}_{[0,L-1]}$ respectively. $\mathbf{v}_i = [\mathbf{v}_i^{(0)}, \mathbf{v}_i^{(1)}]$, $\mathbf{v}_i^{(0)}$ are the information bits and $\mathbf{v}_i^{(1)}$ are the check bits. The PCM of channel codes is shown in Formula (1).

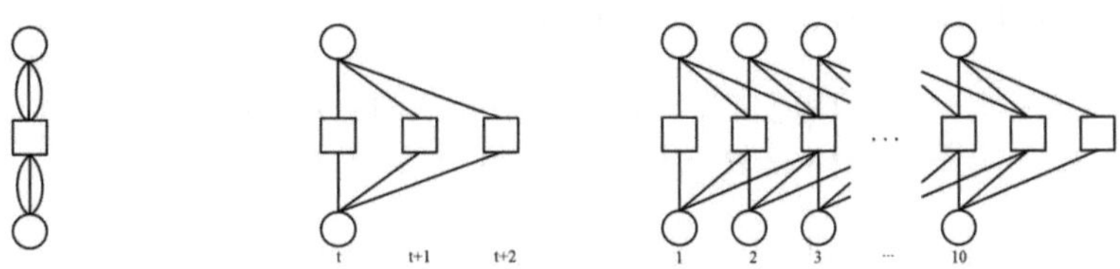

(a)(3,6) regular LDPC protograph (b)the process of edge spreading with m_s=2 (c)protograph of the (3,6,10) SC-LDPC codes

Fig. 2 Construction process of SC-LDPC codes protograph

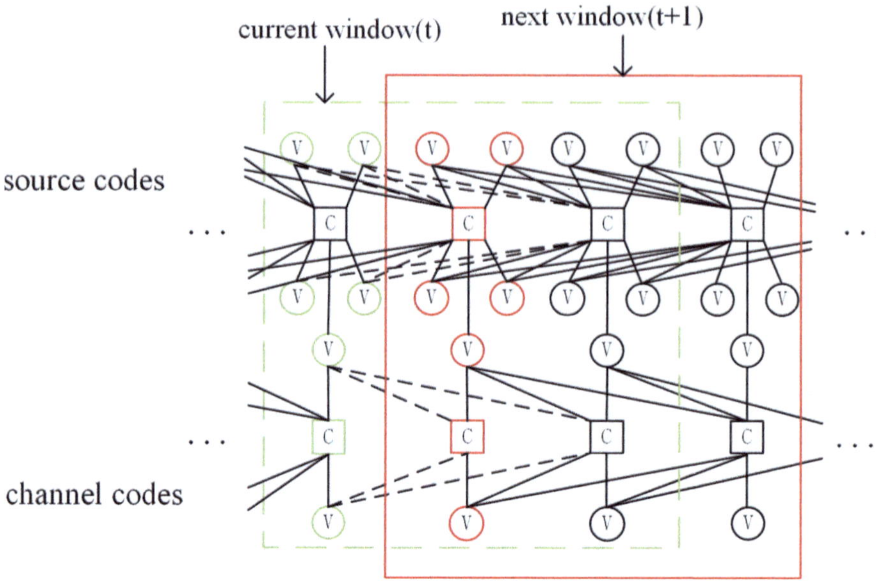

Fig. 3 SC-LDPC codes sliding window decoding algorithm

$$\mathbf{H}^{\mathrm{T}}_{[0,L-1]} = \begin{bmatrix} \mathbf{H}^{\mathrm{T}}_0(0) & & \cdots\ \mathbf{H}^{\mathrm{T}}_{m_s}(0+m_s)\ \mathbf{0} & & & \\ \mathbf{0} & \mathbf{H}^{\mathrm{T}}_0(0) & \cdots & \mathbf{H}^{\mathrm{T}}_{m_s}(1+m_s) & \ddots & \\ \vdots & \mathbf{0} & \ddots\ \vdots & \vdots & \ddots & \\ & & \ddots\ \mathbf{H}^{\mathrm{T}}_0(L-1) & \cdots & & \mathbf{H}^{\mathrm{T}}_{m_s}(L-1+m_s) \end{bmatrix} \tag{1}$$

Information bits $\mathbf{v}_i^{(0)}$ can be counted:

$$v_t^{(j)} = u_t^{(j)},\ j = 1, \ldots, M \tag{2}$$

Partial syndrome $\mathbf{p}_{t,i}$ can be counted:

$$\mathbf{p}_{t,i} = \begin{cases} \mathbf{p}_{t-1,i-1} + \mathbf{v}_{t-1}\mathbf{H}_i^{\mathrm{T}}(t+i-1), & i = 1, \ldots, m_s - 1 \\ \mathbf{v}_{t-1}\mathbf{H}^{\mathrm{T}}_{m_s}(t+m_s-1), & i = m_s \end{cases} \tag{3}$$

So check bits can be computed:

$$\mathbf{v}_t^{(1)} = \boldsymbol{v}_t^{(0)}[\mathbf{H}_0^{(0)}(t)]^{\mathrm{T}} + \mathbf{p}_{t,1} \tag{4}$$

where **H(0) 0**(t) is a $M \times M$ matrix.

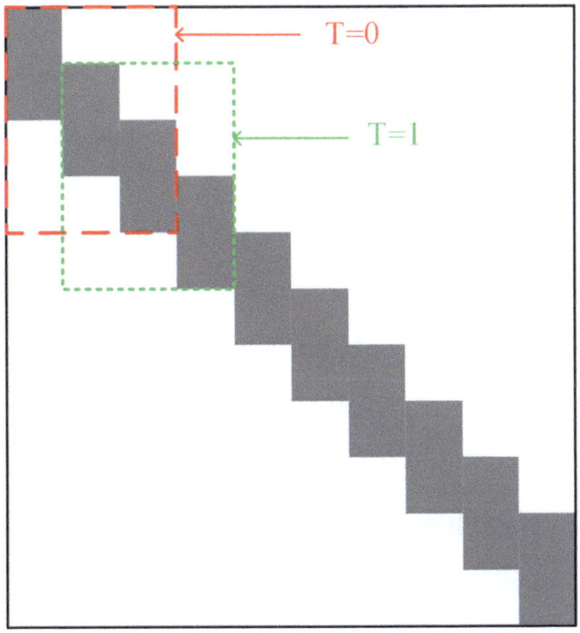

Fig. 4 Sliding window decoding process diagram

2.2 Decoding SC-LDPC Codes in JSCC

Sliding window decoding is to run the BP algorithm in the window to decode the target symbol [11]. Sliding window decoding doesn't need to receive the complete code word, which can reduce the decoding delay. The parity check matrices of SC-LDPC codes presents diagonal band results of non-zero terms, as shown in Fig. 4, so the window can slide down along the parity check matrices. Figure 4 is a schematic diagram with a coupling length of 10 and a window size of 3.

The protograph unit at the left of the window is called the target symbol. When the target symbol decoding is completed, it means that the window decoding is completed. Then save the LLRs of the edge connected to the nodes in the next window (as dotted edges in Fig. 3), and the window slides to the right to translate next target symbol. And variable nodes that overlap the current window with the next window do not need to be initialized when decoding in the next window.

In the JSCC system, there is information interaction between channel coding and source coding. The decoding process is as follows:

Step 1: The log-likelihood ratios (LLRs) of *i-th* variable node of source and channel are Z_i^{sc} and Z_i^{cc} respectively.

$$Z_i^{sc} = \log((1 - p_i)/p_i) \tag{5}$$

$$Z_i^{cc} = 2y_i/\sigma_n^2 \tag{6}$$

where p_i is the probability of 1 in the source, y_i is the code word received by the channel decoder and σ_{n}^2 is the noise variance.

Step 2: Horizontal iterations:

The LLR passed by the check nodes of the source to the variable nodes is

$$\tanh\left(\frac{m_{c,v}^{sc,(k)}}{2}\right) = \tanh\left(\frac{m_v^{cc\to sc,(k)}}{2}\right)\prod_{v'\neq v}\tanh\left(\frac{m_{v',c}^{sc,(k)}}{2}\right), c = 1,\ldots,(L^{sc}+2)M \tag{7}$$

The LLR passed by the check nodes of the channel to the variable nodes is

$$\tanh\left(\frac{m_{c,v}^{cc,(k)}}{2}\right) = \prod_{v'\neq v}\tanh\left(\frac{m_{v',c}^{cc,(k)}}{2}\right), c = 1,\ldots,(L^{cc}+2)M \tag{8}$$

The LLR passed by the check nodes of the source to the variable nodes of the channel is

$$\tanh\left(\frac{m_c^{sc\to cc,(k)}}{2}\right) = \prod_{v'}\tanh\left(\frac{m_{v',c}^{sc,(k)}}{2}\right), c = 1,\ldots,(L^{sc}+2)M \tag{9}$$

Step 3: Vertical iterations:

The LLR passed by the variable nodes of the source to the check nodes is

$$m_{v,c}^{sc,(k)} = Z_v^{sc} + \sum_{c'\neq c} m_{c',v}^{sc,(k-1)}, v = 1,\ldots,4L^{sc}M \tag{10}$$

The LLR passed by the variable nodes of the channel to the check nodes is

$$m_{v,c}^{cc,(k)} = Z_v^{cc} + m_c^{sc\to cc,(k-1)} + \sum_{c'\neq c} m_{c',v}^{cc,(k-1)}, v = 1,\ldots,\frac{1}{2}L^{cc}M \tag{11}$$

$$m_{v,c}^{cc,(k)} = Z_v^{cc} + \sum_{c'\neq c} m_{c',v}^{cc,(k-1)}, v = \frac{1}{2}L^{cc}M+1,\ldots,L^{cc}M \tag{12}$$

The LLR passed by the variable nodes of the channel to the check nodes of the source is

$$m_v^{cc\to sc,(k)} = Z_v^{cc} + \sum_{c'\neq c} m_{c',v}^{cc,(k-1)}, v = 1,\ldots,\frac{1}{2}L^{cc}M \tag{13}$$

Step 4: Judgment:

When the number of iterations reaches the maximum number of iterations or the number of errors is 0, the joint decoding stops, and the LLR of each variable node in the source LDPC codes are calculated according to Formula 14.

$$L(s_v) = Z_v^{sc} + \sum m_{c,v}^{sc,(k)}, v = 1, \ldots, 4L^{sc}M \tag{14}$$

Then estimate the source information bits obtained by decoding:

$$\hat{s} = \begin{cases} 0, & L(s_v) \geq 0 \\ 1, & L(s_v) < 0 \end{cases} \tag{15}$$

3 Analysis of Simulation Result

In this section, the source (3, 12, 16) SC-LDPC codes and the channel (3, 6, 20) SC-LDPC codes are simulated. The probability of 1 in the source is 0.02. The output sequence of channel encoder is modulated by BPSK and then sent to AWGN channel. In the decoding process, the maximum number of iterations is 30, with 20,000 blocks of data per frame.

Figure 5 is a simulation experiment of different spreading factors, where window size is 10. The results show that the larger the spreading factor is, the better the

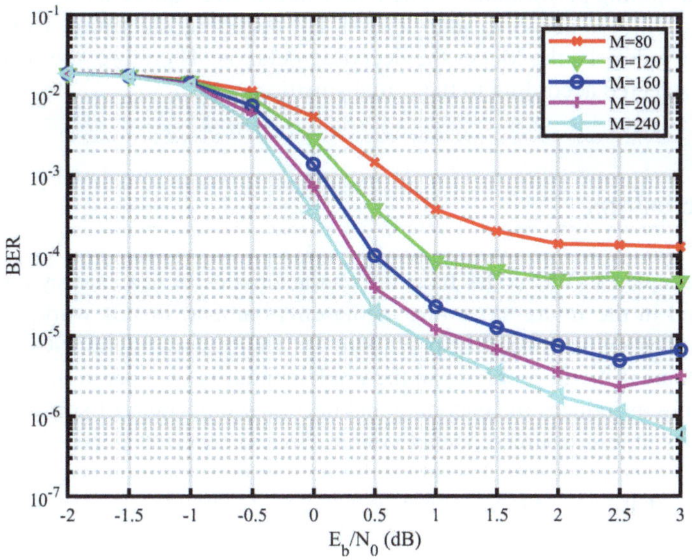

Fig. 5 SC-LDPC codes in JSCC performance of different spreading factors

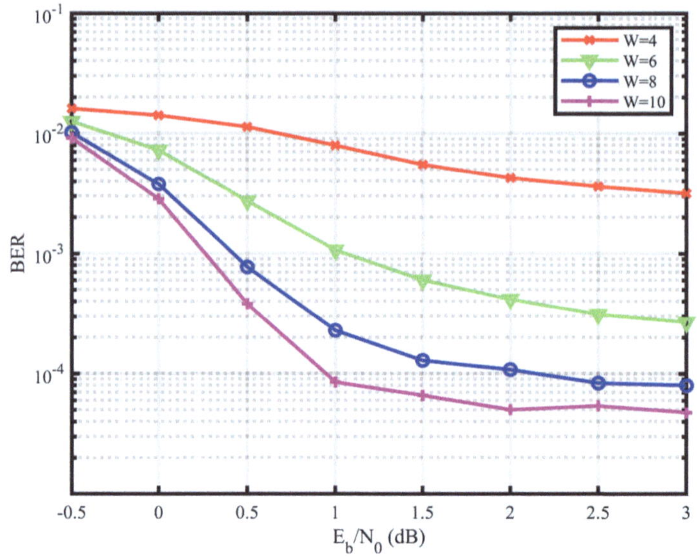

Fig. 6 SC-LDPC codes in JSCC performance of different window

decoding performance is. As we know, the larger the spreading factor is, the longer the code length is. This experimental result is consistent with the conclusion that the longer the code length is, the better the decoding performance is.

Figure 6 is a simulation experiment based on different window sizes, where spreading factor is 120. Experimental results show that the larger the window, the better the decoding performance. When the window size equals the coupling length, the sliding window decoding algorithm is BP decoding algorithm. However, the larger the window is, the longer the decoding delay is. Therefore, the window size should be selected according to the actual situation.

4 Conclusions

Spatially coupled LDPC codes have the characteristics of low latency and will be widely used in the future streaming media environment. In this work, we mainly discuss the coding and decoding process of SC-LDPC code in the JSCC system as there are some difficult problems in the implementation of separation system. This paper mainly introduces the encoding and decoding algorithm of spatially coupled LDPC codes in joint source-channel system. Source encoders generate code words using parity check matrices. Partial syndrome former realization is used in channel encoding so that fewer memory units are used to do encode. In terms of decoding, sliding window decoding is adopted to reduce decoding delay and complexity and there is a message transfer between the source and channel. Simulation results show that the larger the spreading factor, the larger the window, the better the decoding performance. This is consistent with the result that the longer the code length, the better the decoding performance. In the future, we can further search for a better decoding algorithm to improve the decoding performance. What' more, we can design a joint matrix of source and channel to optimize the whole system.

References

1. Sayood, K., Borkenhagen, J.C.: Use of residual redundancy in the design of joint source/channel coders. IEEE Trans. Commun. **39**(6), 838–846 (1991)
2. Pettijohn, B.D., Sayood, K., Hoffman, M.W.: Joint source/channel coding using arithmetic codes. In: Data Compression Conference, pp. 73–82 (2000)
3. Kudekar, S., Richardson, T.J., Urbanke, R.L.: Threshold saturation via spatial coupling: why convolutional LDPC ensembles perform so well over the BEC. IEEE Trans. Inf. Theory **57**(2), 803–834 (2011)
4. Jimenez, F.A., Zigangirov, K.S.: Time-varying periodic convolutional codes with low-desity parity-check matrix. IEEE Trans. Inf. Theory **45**(6), 2181–2191 (1999)
5. Lentmaier, M., Sridharan, A., Costello, D.J., et al.: Iterative decoding threshold analysis for LDPC convolutional codes. IEEE Trans. Inf. Theory **56**(10), 5274–5289 (2010)
6. Klaiber, K., Cammerer, S., Schmalen, L., et al.: Avoiding burst-like error patterns in windowed decoding of spatially coupled LDPC Codes. In: IEEE 10th International Symposium on Turbo Codes & Iterative Information Processing, pp. 1–5 (2018)
7. Mo, S., Chen, L.: Improved sliding window decoding of spatially coupled low-density parity-check codes. In: IEEE Information Theory Workshop, pp. 126–130 (2017).
8. Pusane, A.E., Feltstrom, A.J., Sridharan, A., et al.: Implementation aspects of LDPC convolutional codes. IEEE Trans. Commun. **56**(7), 1060–1069 (2008)
9. Iyengar, A.R., Papaleo, M., Siegel, P.H., et al.: Windowed decoding of protograph-based LDPC convolutional codes over erasure channels. IEEE Trans. Inf. Theory **58**(4), 2303–2320 (2012)
10. Golmohammadi, A., Mitchell, D.G.M.: Concatenated spatially coupled LDPC codes for joint source-channel coding. In: IEEE International Symposium on Information Theory (ISIT), pp. 631–635 (2018)
11. Yamei, Z., Lin, Z., Chen, C., et al.: Sliding window decoding of the spatially coupled LDPC code over Rayleigh fading channels **47**(6), 78–83 (2020)

Simplified Design of Multilevel Coding Scheme with Polar Codes for High-Order Modulation

Lei Xu, Lin Zhou, Lin Wang, Congyi Wang, and Sanya Liu

Abstract Polar codes are a type of forward error correction (FEC) codes that can achieve the capacity of a discrete memoryless symmetric channel. Polar codes are accepted by the 5G standard due to the excellent error correction performance. The next-generation mobile communication system requires higher spectrum efficiency, so it is significant to study the easy-to-implement and efficient coded modulation scheme of polar codes for high-order modulation. The commonly used polar coded modulation schemes in communication systems mainly include bit-interleaved coded modulation (BICM) and multilevel coding (MLC). Aiming at the requirements of higher spectrum efficiency and higher reliability of future communication systems, in this work, we propose an improved MLC scheme based on polar codes to reduce the difficulty of MLC scheme design. Since the quadrature amplitude modulation (QAM) can be decomposed into two orthogonal pulse amplitude modulation (PAM) in the MLC system, we performed simulations based on PAM in an additive white Gaussian noise (AWGN) channel. Compared with the conventional BICM scheme, the proposed MLC scheme provides performance gains of 0.4 dB and 1.48 dB under 8-PAM and 16-PAM, respectively. At the same time, while the proposed MLC scheme provides a large amount of performance gain at 16-PAM, the complexity of the soft-decision decoder of the entire system is reduced by 55%.

1 Introduction

Polar codes are a class of error-correcting codes proposed by Arikan in 2009 based on channel polarization theory, and they were theoretically proven to achieve symmetric channel capacity on binary-input discrete memoryless channels (B-DMC) at infinite

L. Xu · L. Zhou (✉) · C. Wang · S. Liu
Xiamen Key Laboratory of Mobile Multimedia Communications, College of Information Science and Engineering, National Huaqiao University, Xiamen 361021, Fujian, China
e-mail: linzhou@hqu.edu.cn

L. Wang
China Mobile Communications Group Xiamen Co., Ltd. Xiamen, Fujian 361000, China

L. C. Jain et al. (eds.), *Advances in Wireless Communications and Applications*,
Smart Innovation, Systems and Technologies 299,
https://doi.org/10.1007/978-981-19-2255-8_8

code length [1]. Polar codes, as a new star in the field of channel coding, have received much attention from academia and industry since it was proposed, and became the coding scheme for the control channel in the 5G mobile communication standard enhanced Mobile BroadBand (eMBB) scenario in 2016.

The coded modulation (CM) problem has been a very popular research topic, and the early studies of polar codes were mainly based on binary channels, where all coded bits are transmitted through the same B-DMC and then decoded by the receiver. However, there are some problems in applying polar code directly to higher-order modulation. The construction of polar codes is based on Binary Phase Shift Keying (BPSK) to select a good (very high reliability) sub-channel for transmitting information bits based on the sub-channel reliability. In this case, the direct application of the polar codes constructed for independent and identical distribution will cause a reliability mismatch problem, resulting in performance loss. The combination of polar codes and CM is an important and meaningful research direction [2], and usually, polar coded modulation schemes include multilevel polar coded modulation [3] and bit interleaved polar coded modulation [4].

Among them, the MLC scheme is optimal from the point of view of information theory, because it is completely based on the chain rule of mutual information, which protects bit levels of different reliability in groups, and approaches channel capacity through multi-stage decoding [5]. However, the design of the MLC scheme requires multiple component codes, and the system will be very complicated for high-order modulation. The BICM scheme counteracts interference by introducing an interleaver to discrete burst errors into random errors in the channel [6]. The BICM scheme is endeared because of its simplicity and flexibility. Nevertheless, the performance gain of the BICM scheme under high-order modulation and high spectral efficiency is limited, which is far inferior to the MLC scheme [7].

Recently, another MLC scheme of concatenated codes for high-order modulation was introduced in [8, 9], in which the MLC scheme is designed to be of low-complexity and offers better performance over BICM. Inspired by this, considering that some bit levels in the MLC system are already very reliable at high code rates, we separate these bit levels from the other bit levels and utilize polar codes to separately encode the less reliable bit levels as least significant bits (LSBs), while the other bit levels as the most significant bits (MSBs) are not encoded. We apply the strong error correction capability of the polar codes to ensure the reliability of the bit levels in the LSBs, at the same time, the encoding code length is reduced by using the split layer scheme, which brings the reduction of system power consumption. The simulation results show that the proposed MLC scheme outperforms the BICM scheme.

2 Polar Codes

2.1 Channel Polarization

Channel polarization is the core theory of polar codes, which consists of channel combining and channel splitting. The phenomenon of channel polarization is that when the code length N tends to infinity, a set of independent binary input channels will be transformed into two types of extreme channels with symmetric capacity approaching 0 or 1, namely, a noiseless channel and a full noise channel. We transmit the information bits on the noiseless channel, and transmit the information agreed by the sender and receiver in advance on the full noise channel, that is, the frozen bits.

2.2 Polar Encoding

Polar Codes apply a generator matrix $\mathbf{G}_N = \mathbf{B}_N\ \mathbf{F}^{\otimes n}$ for coding, where $N = 2^n$ is the code length, "$\otimes$" is the Kronecker product, $\mathbf{F}$ represent the kernel matrix:

$$\mathbf{F} = \begin{bmatrix} 1 & 0 \\ 1 & 1 \end{bmatrix} \tag{1}$$

and $\mathbf{B}_N$ is the permutation operation:

$$\mathbf{B}_N = \mathbf{R}_N\left(\mathbf{I}_2 \otimes \mathbf{B}_{N/2}\right) \tag{2}$$

where $\mathbf{I}_2$ represent an identity matrix, $\mathbf{R}_N$ is the permutation matrix.

The encoding operation of polar codes (N, K) is divided into two steps. Firstly, a good set of sub-channels are selected to carry K information bits through the method of coding construction, and the remaining sub-channels carry frozen bits to obtain the message sequence $\boldsymbol{u}_1^K$. In this work, we choose the polarization weight method proposed by HUAWEI [10]. Then, the message sequence $\boldsymbol{u}_1^K$ and the generator matrix $\mathbf{G}_N$ are multiplied to obtain the coded sequence $varvecx_1^N$:

$$\boldsymbol{x}_1^N = \boldsymbol{u}_1^N \mathbf{G}_N \tag{3}$$

2.3 Polar Decoding

Corresponding to the channel splitting, Arikan proved that when the code length of polar codes is large enough, a better asymptotic performance can be obtained by applying the successive cancellation (SC) decoding algorithm. SC decoding is a single-pass algorithm where the decoding process of the ith bit requires the use of the log-likelihood ratio (LLR) of the channel output and the decoding result of the previous $(i-1)$th bits, and upon receiving them, computes the LLR

$$L_N^{(i)}(y_1^N, \hat{u}_1^{i-1}) = \ln \frac{W_N^{(i)}(y_1^N, \hat{u}_1^{i-1}|u_i = 0)}{W_N^{(i)}(y_1^N, \hat{u}_1^{i-1}|u_i = 1)} \tag{4}$$

and makes its decision as

$$\hat{u}_i = \begin{cases} 0, \ L_N^{(i)}(y_1^N, \hat{u}_1^{i-1}) \geq 0 \\ 1, \ L_N^{(i)}(y_1^N, \hat{u}_1^{i-1}) < 0 \end{cases} \tag{5}$$

the complexity of SC decoding mainly comes from the calculation of LLR, which can be calculated by parity grouping recursively

$$L_{2N}^{(2i-1)}\left(y_1^{2N}, \hat{u}_1^{2i-2}\right) = \ln \frac{\mathrm{e}^{L_1+L_2}+1}{\mathrm{e}^{L_1}+\mathrm{e}^{L_2}} \tag{6}$$

$$L_{2N}^{(2i)}\left(y_1^{2N}, \hat{u}_1^{2i-1}\right) = \left(1-2\hat{u}_{2i-1}\right)L_1 + L_2 \tag{7}$$

L_1 and L_2 were defined as:

$$\begin{cases} L_1 = L_N^{(i)}\left(y_1^N, \widehat{u}_{1,o}^{2i-2} \oplus \widehat{u}_{1,e}^{2i-2}\right) \\ L_2 = L_N^{(i)}\left(y_{N+1}^{2N}, \widehat{u}_{1,e}^{(2i-2)}\right) \end{cases} \tag{8}$$

$\hat{u}_{1,\mathrm{o}}^{2i-2}$ denotes the odd grouping $\left(\hat{u}_1, \hat{u}_3, \ldots, \hat{u}_{2i-3}\right)$ of $\hat{u}_i$ and $\hat{u}_{1,\mathrm{e}}^{2i-2}$ denotes the even grouping $\left(\hat{u}_2, \hat{u}_4, \ldots, \hat{u}_{2i-2}\right)$ of $\hat{u}_i$. The recursive process starts with the reception of the channel LLR $L_1^{(1)}(y_i)$ and keeps recursion to finally get $L_N^{(i)}(y_1^N, u_1^{i-1})$.

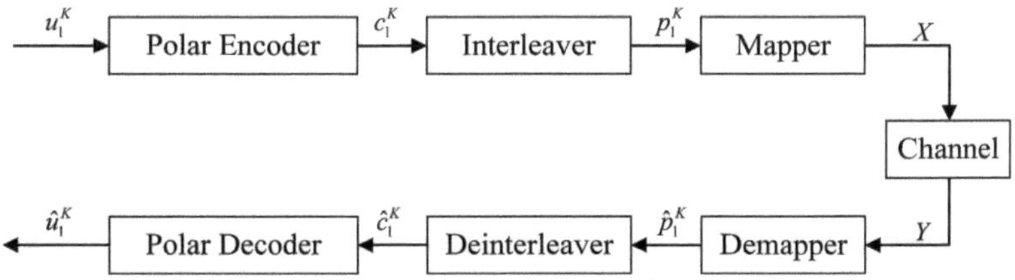

Fig. 1 The BICM scheme model

3 BICM Scheme and the Proposed MLC Scheme

3.1 BICM Scheme

BICM is a practical coded modulation scheme, which is widely used in engineering practice. The purpose is to use an interleaver to disrupt the correlation of the coded bit sequence and to disperse the bits with continuous errors into different constellation symbols, so as to obtain a larger codeword dispersion. The BICM scheme of the Polar codes is a simple and practical high-order coded modulation scheme. The design of the interleaver is the focus, we use different interleaving rules to disrupt the coded bit sequence and discretize burst errors, thereby improving system performance. Figure 1 is a bit-interleaved polar coded modulation scheme model. Compared with the high-order modulation system in which coding and modulation are independent of each other, an interleaver is added between the encoder and the mapper, and a de-interleaver is added correspondingly at the receiver. This reflects the advantages of the BICM. The advantage of the BICM scheme is its low complexity, but it is not optimal in information theory.

3.2 The Proposed MLC Scheme

Let A denote the set of signal constellation points, and its cardinality given by $|A| = 2^m$. Let $X \in A$ denote the transmit symbol composed of codewords and Y is the receive signal. Then, the channel can be equivalent to m parallel sub-channels according to the mutual information chain rule:

$$\begin{aligned} I(Y;X) &= I\big(Y;x^{(1)},\ldots,x^{(m)}\big) \\ &= I\big(Y;x^{(1)}\big) + I\big(Y;x^{(2)}|x^{(1)}\big) + \cdots + I\big(Y;x^{(m)}|x^{(1)},\ldots,x^{(m-1)}\big) \end{aligned} \quad (9)$$

the capacity C^i of the equivalent sub-channel is given by the respective mutual information:

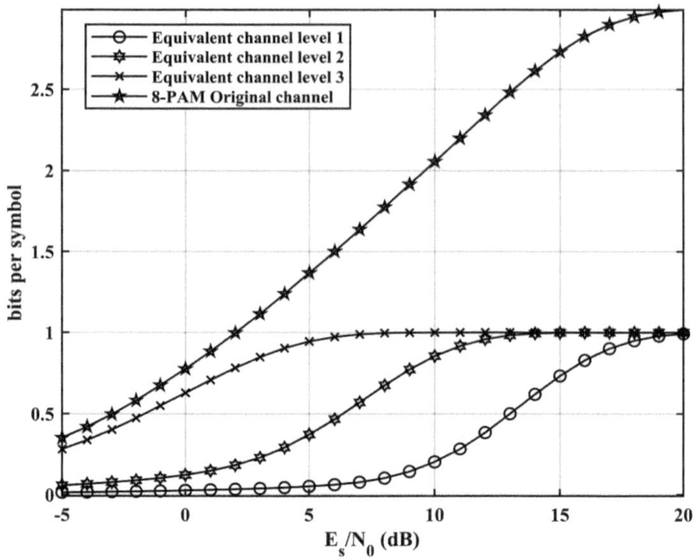

Fig. 2 The capacity of equivalent channels with 8-PAM

$$
\begin{aligned}
C^i &= I\left(Y; x^{(m)} | x^{(1)}, \ldots, x^{(m-1)}\right) \\
&= \mathrm{E}_{x^0 x^1 \ldots x^{i-1}}\left\{C\left(A\left(x^0 x^1 \ldots x^{i-1}\right)\right)\right\} - \mathrm{E}_{x^0 x^1 \ldots x^i}\left\{C\left(A\left(x^0 x^1 \ldots x^i\right)\right)\right\}
\end{aligned} \tag{10}
$$

where $C\left(A\left(x^0 x^1 \ldots x^i\right)\right)$ denotes the coded modulation capacity, and E represents an expectation operation. For example, Fig. 2 shows the channel capacity of each bit level using Ungerboeck partitioning with 8-PAM modulation over the AWGN channel.

It is important to assign code rates to each level according to specific rules, and the capacity rule is one of the effective design rules in traditional MLC schemes. However, the capacity C^i of the higher bit levels of the MLC scheme tends to 1 at high spectral efficiency, when still coding these bit levels for protection brings less benefit and generates a lot of unnecessary system overhead.

Therefore, we simplify the MLC scheme to the two-level scheme shown in Fig. 3, bind the bit level with capacity C^i tending to 1 as the MSBs and the other bit levels as the LSBs. Then, the LSBs are encoded with a polar encoder, while the MSBs are not encoded. At the receiver, the input Y is divided into MSBs and LSBs, the LSBs are demodulated to obtain LLR, which is sent to a polar decoder to obtain the decoding result $\hat{\boldsymbol{u}}^{(2)}$ of LSBs. After that, $\hat{\boldsymbol{u}}^{(2)}$ will be applied to assist the hard-decided demodulator of MSBs to obtain the decision result $\hat{\boldsymbol{u}}^{(1)}$.

We describe the subset partitioning and constellation labeling rules as follows. We divide the constellation signal point label into two subsets: MSBs and LSBs, and they are divided according to Ungerboeck partitioning. Meanwhile, Gray code is used for labeling inside the subset. For example, for an 8-PAM constellation consisting of

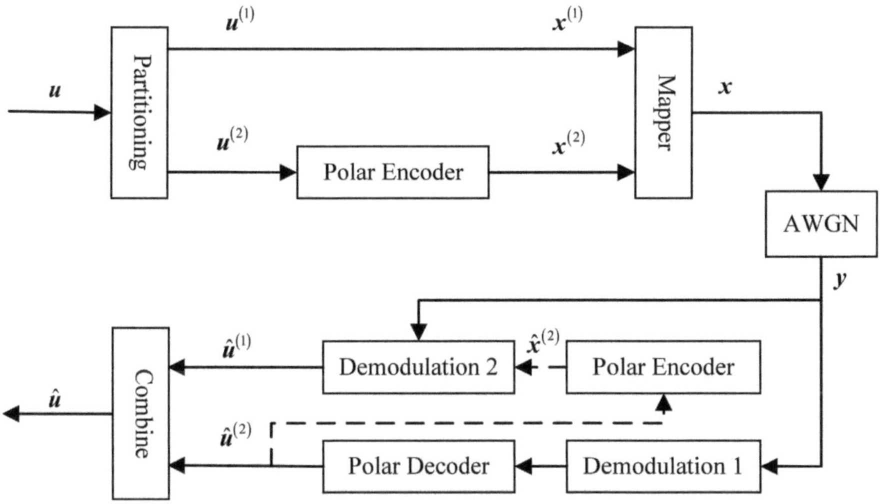

Fig. 3 The proposed MLC scheme model

three bits per constellation point, we use the least reliable bit level as the LSB and the remaining two bit levels as the MSBs. We apply the following labeling for the 8-PAM constellation: $A_{8-\mathrm{PAM}} = (000, 010, 110, 100, 001, 011, 111, 101)$. Through this labeling, it is possible to clearly distinguish the reliability between the LSB and the MSBs, thus making the MSBs channel sufficiently reliable.

4 Simulation Results and Analysis

In this section, we show the simulation results to compare the proposed MLC scheme and BICM scheme under AWGN channel. We performed simulations with 8-PAM and 16-PAM modulation. In the 8-PAM scheme, the spectral efficiency β is 2.5 bits per symbol, and the code length N is 1536 for the proposed MLC scheme and 2046 for BICM (obtained using punching). For 16-PAM scheme, β is 3 bits per symbol and N is 1024 are used for simulation. In addition, the last two bit-levels are selected to the MSBs, because their reliability is high enough.

As shown in Fig. 4, for the proposed MLC scheme, the performance gains are 0.4 dB and 1.48 dB under 8-PAM and 16-PAM modulation compared to BICM at BER = 10–5. From the simulation results, it can be seen that the proposed scheme has higher performance gain at higher modulation orders. At the same time, in the proposed MLC scheme, since only LSBs are encoded and decoded, the number of encoders and decoders is partially decreased, which reduces the complexity of the system. In the polar coded modulation scheme, the system complexity mainly comes from the soft-decision decoder of polar codes, which involves a large number of floating-point operations. Therefore, we can define the relative complexity reduction

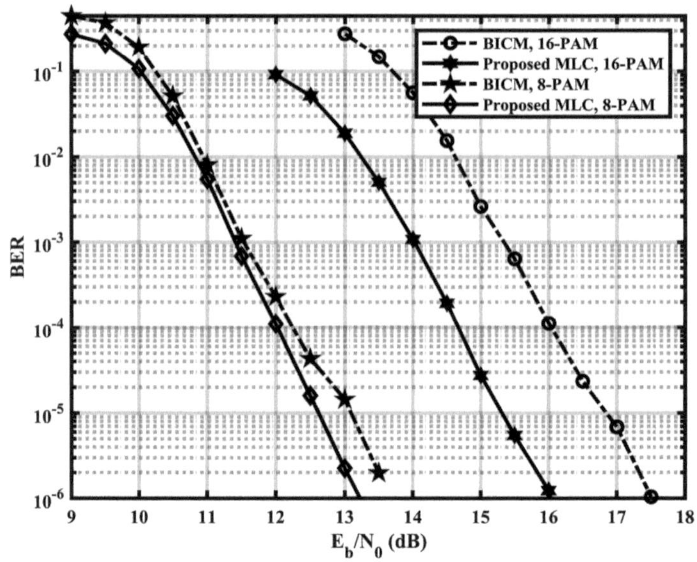

Fig. 4 BER performance of the proposed MLC scheme and BICM

of the entire system as the complexity reduction of the decoder in the system. Since the decoding complexity of the SC decoder is $O(N \log_2 N)$, the relative complexity reduction of the proposed MLC scheme to BICM scheme can be calculated as:

$$T_{\text{reduce}} = 1 - \frac{T_{\text{MLC}}}{T_{\text{BICM}}} \tag{11}$$

T_{MLC} denotes the decoding complexity of the proposed MLC scheme and T_{BICM} denoted BICM decoding complexity. For 16-PAM, the proposed MLC scheme offers 55% reduction of SC decoder. Compared with the difficult implementation of the traditional MLC scheme and the poor performance of the BICM system, the proposed scheme achieves a compromise between performance and complexity.

5 Conclusion

Facing the requirements of high spectral efficiency and high-order modulation in future communication systems, in this article, we propose an improved two-level MLC scheme that can be applied when the code rate is high enough. To reduce the difficulty of implementing the MLC scheme, the structure is redesigned to have only two levels, in which LSBs and MSBs are divided according to Ungerboeck partitioning, and gray labeling is used inside the set to maximize the reliability difference between them. In this case, the bits in the LSBs are protected by polar coding and the remaining bits are transmitted through the MSBs channel and remain uncoded state. The simulation analysis was performed in two scenarios of 8-PAM and 16-PAM over AWGN channel. The results show that the proposed MLC scheme is better than the BICM scheme in the 5G standard, and the higher the modulation order, the greater the performance gain. In addition, since only the LSBs are coded, the overall code length of the system is decreased, resulting in a reduction in the complexity of the system. The proposed scheme is very effective with high baud rate, and how to use the scheme to obtain the ideal performance with low rate is a problem that we need to consider in our future work. In addition, we only consider one-dimensional modulation at present, and we will optimize the proposed scheme under higher-dimensional modulation in the future.

References

1. Arikan, E.: Channel polarization: a method for constructing capacity-achieving codes for symmetric binary-input memoryless channels. IEEE Trans. Inf. Theory **55**(7), 3051–3073 (2009)
2. Seidl, M., Schenk, A., Stierstorfer, C., Huber, J.B.: Polar-Coded modulation. IEEE Trans. Commun. **61**(10), 4108–4119 (2013)
3. Zhang, Q.S., Liu, A.J., Pan, X.F., Zhang, Y.X.: Symbol-based belief propagation decoder for multilevel polar coded modulation. IEEE Commun. Lett. **21**(1), 24–27 (2017)
4. Afser, H., Tirpan, N., Delic, H., Koca, M.: Bit-interleaved polar-coded modulation. In: 2014 IEEE Wireless Communications and Networking Conference (WCNC), pp. 480–484. IEEE, Istanbul, Turkey (2014)
5. Wachsmann, U., Fischer, R.F.H., Huber, J.B.: Multilevel codes: theoretical concepts and practical design rules. IEEE Trans. Inf. Theory **45**(5), 1361–1391 (1999)
6. Tian, K., Liu, R., Wang, R.: Joint successive cancellation decoding for bit-interleaved polar coded modulation. IEEE Commun. Lett. **20**(2), 224–227 (2016)
7. Bisplinghoff, A., Langenbach, S., Kupfer, T.: Low-power, phase-slip tolerant, multilevel coding for M-QAM. J. Lightwave Technol. **35**(4), 1006–1014 (2017)
8. Mehmood, T., Yankov, M.P., Iqbal, S., Forchhammer, S.: Flexible multilevel coding with concatenated polar-staircase codes for M-QAM. IEEE Trans. Commun. **69**(2), 728–739 (2021)
9. Barakatain, M., Lentner, D., Becherer, G., Kschischang, F.R.: Performance-complexity trade-offs of concatenated FEC for higher-order modulation. J. Light-wave Technol. **38**(11), 2944–2953 (2020)
10. Zhou, Y., Li, R., Zhang, H., Luo, H., Wang, J.: Polarization weight family methods for polar code construction. In: Proceedings of 2018 IEEE 87th Vehicular Technology Conference (VTC Spring), pp. 1–5. IEEE, Porto, Portugal (2018)

Performance Analysis of Probabilistic Shaped Polar Code in 5G

Congyi Wang, Lin Zhou, Lei Xu, Qiwang Chen, and Yidong Ke

Abstract As a coding scheme in the scenario of enhanced mobile bandwidth in 5G communication, polar codes have been deeply studied due to its communication reliability. Probability shaping is an important method to approach the channel capacity limit. Therefore, it is necessary to study the combination of polar codes and probabilistic shaping. In this paper, we analyze the performance of the shaping systematic polar codes (SPC) and the shaping non-systematic polar codes (NSPC) in SC decoding algorithms. It is proposed to combine systematic polar codes with probabilistic shaping technology, hoping to induce a sub-optimization symbol probability distribution. We found that in the scheme of probabilistic shaping based on the source sequence, the non-systematic polar code encoding method will force the bit probability in the encoded codeword to be equal, which will greatly affect the performance gain brought by the probability shaping. Therefore, in this paper, we use systematic polar encoder to protect the unequal probability characteristics of 0 and 1 in the source sequence, so that the distribution of the channel input symbols tends to the Maxwell Boltzmann distribution and achieve a performance gain of 0.6 dB.

1 Introduction

Arikan first proposed polar codes in 2008 [1]. Polar codes are proven to be a channel coding scheme that can approach the symmetric capacity of binary-input discrete memoryless channel (B-DMC). Afterwards, Arikan proposed the coding method of systematic polar codes and pointed out that the performance of systematic polar codes in bit error rate (BER) is better than NSPC [2]. Because polar codes have lower

C. Wang · L. Zhou (✉) · L. Xu · Q. Chen
Xiamen Key Laboratory of Mobile Multimedia Communications, College of Information Science and Engineering, National Huaqiao University, Xiamen 361021, Fujian, China
e-mail: linzhou@hqu.edu.cn

Y. Ke
QLX (Xiamen) Technology Co., Ltd, , Xiamen 361000, Fujian, China

L. C. Jain et al. (eds.), *Advances in Wireless Communications and Applications*,
Smart Innovation, Systems and Technologies 299,
https://doi.org/10.1007/978-981-19-2255-8_9

coding and decoding complexity and excellent codeword error correction performance, polar codes have been chosen as the channel coding scheme in 5G enhanced mobile broadband (eMBB) scenarios in recent years.

Many scholars have conducted in-depth research on probabilistic shaping techniques in NSPC. A probabilistic shaping scheme that uses a polar decoder as a shaping code generator is proposed in [3]. This scheme can avoid additional shaping code generators and at the same time avoid the addition of corresponding shaping decoders at the receiving end. The dynamic freeze bit is proposed in [4], which combines the non-systematic polarization code and the many-to-one mapper to generate an ideal channel input symbol distribution. The random number generator uses a recoverable random number seed to generate a sequence of random numbers and put into the frozen bit index. At the receiving end, these pseudo-random numbers are used as prior information to eliminate the ambiguity introduced by the many-to-one mapping. A many-to-one probability shaping technique based on low-density parity-check codes (LDPC) was sent to the fiber channel for experimental simulation in [5]. Experiments show that a nearly Gaussian PAM8 signal obtained from a uniformly distributed PAM16 achieves superior receiver power sensitivity and excellent transmit fiber power optimization in the optical fiber network.

Many scholars have studied the probabilistic shaping with NSPC in depth, but there are not many studies on the probability shaping of systematic polar codes. In this paper, we first describe the systematic and non-systematic coding methods of polar codes, and then introduce probabilistic shaping of the source. The experiment simulates the probability distribution state of the signal points after different coding methods, in order to observe the influence of the systematic polarization codes and the non-systematic polarization codes on the probabilistic shaped source. Finally, we experimentally simulated the performance of the systematic shaped polar codes and the non-systematic shaped polar codes under the SC decoding algorithm, and compared them with the performance of uniformly distributed signal points.

2 Related Work

2.1 Non-Systematic Polar Coding

Polar codes are a kind of (N, K) linear codes, N represent the length of the codeword and K represent the number of information bits. The channel is divided into N virtual sub-channels. Through the Gaussian approximation method and the polarization weight method, K sub-channels with the highest reliability can be selected to transmit information bits, the remaining $N - K$ sub-channels are used to transmit fixed bits known to both the receiver and the transmitter. Fixed bits are called frozen bits.

For a binary source sequence u with code length N, the non-systematic codewords x can be obtained by

$$x = uG_N \tag{1}$$

where x is the codewords, G_N represent a $N \times N$ matrix. The generator matrix G_N can be expressed as

$$G_N = B_N F^{\otimes n}, \quad F = \begin{bmatrix} 1 & 0 \\ 1 & 1 \end{bmatrix} \tag{2}$$

where B_N represents a column permutation matrix, which separates the odd and even columns in the matrix. $F^{\otimes n}$ represents the Kronecker power.

The source sequence u can be divided into u_A, $(u_A = u_k)$ and u_{A^c}, $(u_{A^c} = 0)$, where A contains the index of K sub-channels with the highest reliability, and A^c represents the index set of the remaining $N - K$ sub-channels. Therefore, (1) can be written as

$$x = u_k G_A \tag{3}$$

where G_A is the sub-matrix contains the row of index A in G_N.

2.2 Systematic Polar Coding and SC Decoding

The systematic polar encoding divides generator matrix G_N into multiple sub-matrix and codeword x is divided into x_A and x_{A^c}. The systematic polar codewords x_A and x_{A^c} are obtained by formula (4).

$$\begin{aligned} x_A &= u_A G_{AA} + u_{A^c} G_{A^c A} \\ x_{A^c} &= u_A G_{AA^c} + u_{A^c} G_{A^c A^c} \end{aligned} \tag{4}$$

where G_{AA^c} represents a submatrix composed of rows contained in set A and columns contained in set A^c in G_N.

When G_{AA} is an invertible matrix, there is a systematic encoder, u_A can be calculated by the following formula (5)

$$u_A = (x_A - u_{A^c} G_{A^c A})(G_{AA})^{-1} \tag{5}$$

then x_{A^c} can be obtained through (4) and (5). A new systematic polar coding method with lower hardware occupancy resources is proposed in [6]. Compared with the method in [1], this method requires less calculation.

SC decoding is an algorithm proposed in [1], which divides the received bit likelihood ratio (LR) sequence into two parts, A and A^c. For the received ith bit LR, the frozen bit $\hat{u}_i$ can be directly hard-decided to 0 when $i \in A^c$. When $i \in A$, the SC decoder calculates the LR of the ith codeword $\hat{u}_i$ according to the decision result of all $i - 1$ bits before $\hat{u}_i$.

$$L_N^i(y_1^N, \hat{u}_1^{i-1}) = \frac{P_N^i(y_1^N, \hat{u}_1^{i-1}|u_i = 0)}{P_N^i(y_1^N, \hat{u}_1^{i-1}|u_i = 1)} \tag{6}$$

where y_1^N represents the channel output vector, $\hat{u}_1^{i-1}$ represents the decision result of $i - 1$ bits, and P_N is the channel transition probability. Then, make a hard decision on the LR value to obtain a bit estimate $\hat{u}_i$.

$$\hat{u}_i = \begin{cases} 0, & \text{if } L_N^i(y_1^N, \hat{u}_1^{i-1}) \geq 1 \\ 1, & \text{if } L_N^i(y_1^N, \hat{u}_1^{i-1}) < 1 \end{cases} \tag{7}$$

The decoder keeps the judgment results obtained and uses it to help calculate the LR of all bits after the ith bit. Due to the problem of error propagation in SC decoding, a decoding method that preserves multiple sdecoding paths is proposed in [7].

2.3 *Probabilistic Shaping*

For traditional communications, the channel input symbol distribution is uniformly distributed. However, this is not optimal for most channels. For example, the optimal symbols probability distribution of the additive white Gaussian noise (AWGN) channel is the Maxwell Boltzmann distribution, that is, the MB distribution. Therefore, the uniform distribution of channel input symbols will bring performance loss to the system, which we call shaping loss. To solve this problem, probabilistic shaping technology has been introduced into channel coding. Probabilistic shaping increases the frequency of low-energy signal points, reduces the frequency of high-energy points, and reduces the average power. This paper introduces the probabilistic shaping technique proposed in [8].

3 Probabilistically Shaped Systematic Polar Code

A structure combining probability shaping and channel coding in a single-layer BICM is proposed in [9]. We use the structure in this paper to combine the systematic polar code with probabilistic shaping, and its structure is shown in Fig. 1. The difference from the literature is that in the structure adopted in this paper, the probability shaping does not depend on the mapper, but converts the binary equal-probability source sequence to the non-equal-probability source before encoding.

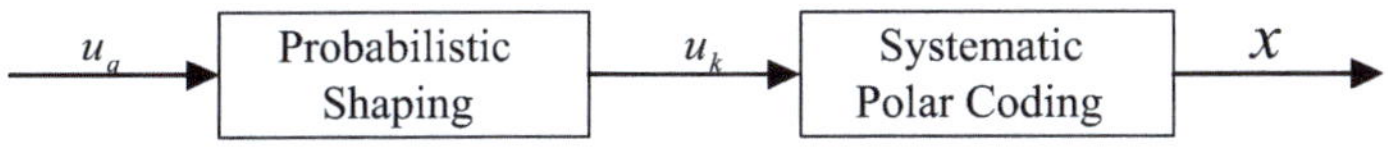

Fig. 1 Probabilistic shaping structure of systematic polar codes

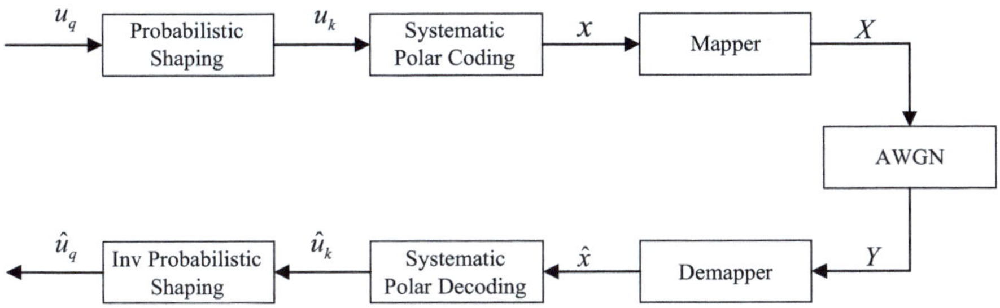

Fig. 2 Single-layer shaped polar codes scheme

Due to the limitation of the generation matrix of traditional systematic polar coding, we are adopting a kind of systematic polar coding scheme consisting of two-step non-systematic coding [10]. The system frame diagram is shown in Fig. 2.

At the transmitting end, a binary uniformly distributed bit source u_q of length q is converted into a non-equal probability binary vector u_k of length k by the shaping code encoder. The vector u_k is sent to a systematic polar encoder with a code rate of k/N to generate a code vector x, which is then mapped to 2^m-PAM signal points by a mapper. The PAM symbol sequence X is sent into the AWGN channel.

At the receiving end, the channel output symbol Y is sent to the de-mapper to calculate the log-likelihood ratio (LLR) of the encoded codeword x, and the source bit estimate $\hat{u}_k$ is calculated through a systematic polar decoder, and then it is sent to the shaping code decoder.

4 Experiment Result and Data Analysis

In this part, we simulate and analyze the influence of the two encoding methods of the polar code on the probabilistically shaped source sequence. At the same time, the performance of probabilistic shaped systematic polar codes and probabilistic shaped NSPC under same code rates is compared. The influence of systematic polar coding and non-systematic polar coding on probability shaping is shown in Fig. 3. It can be seen that the NSPC eliminates the unequal probability characteristics of 0 and 1 in the formed source sequence, resulting in a uniform distribution of the mapped PAM4 signal points. Compared with NSPC, systematic polar codes greatly remain the "0" and "1" bit unequal probability characteristics brought about by probabilistic shaping. Therefore, the combination of system polar coding and source probability shaping

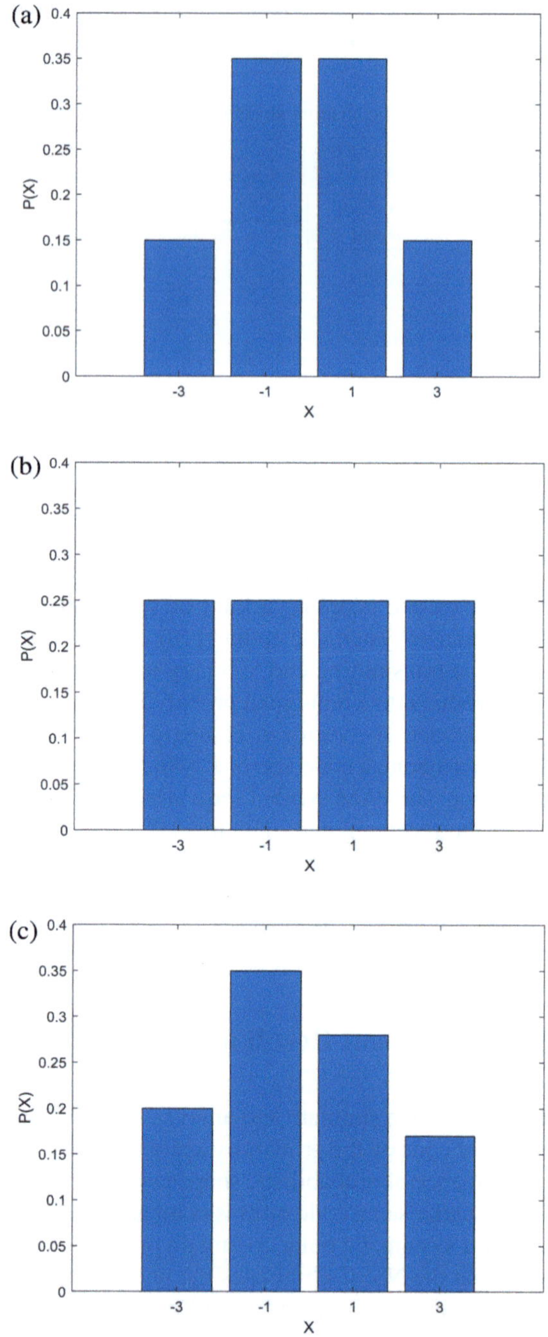

Fig. 3 **a** is the shaped PAM4 symbol probability distribution without polar coding. **b, c** are the symbol probability distributions of probability-shaped non-systematic polarization codes and probability-shaped systematic polarization codes after PAM4 mapping

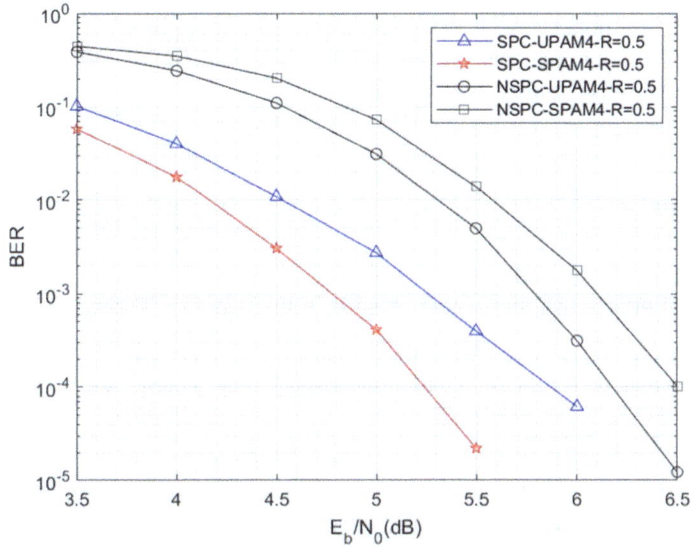

Fig. 4 Performance of PS-SPC and PS-NSPC under SC decoding algorithm

can form a constellation point probability distribution with progressive standard MB distribution.

Therefore, this paper adopts a scheme that combines probabilistic shaping with systematic polar codes, and retains the shaping gain brought by probability shaping. The corresponding performance of this scheme is shown in Fig. 4.

Compared with the uniformly distributed systematic polar codes with code length $N = 1024$ and code rate $R = 0.5$, the systematic polar codes after probabilistic shaping have better BER performance. The shaping gain reaches about 0.6 dB when $\text{BER} = 10^{-4}$. The NSPC eliminate the shaping gain brought by the probabilistic shaping. At the same time, because probabilistic shaping introduces $k - q$ redundant bits before encoding, resulting in a loss of code rate, the performance is worse than that of uniformly NSPC.

5 Conclusions

Systematic polar codes are superior to NSPC in terms of bits error rate. The combination of probability shaping and systematic polar codes is an important means to approximate the channel capacity limit. Therefore, it is very meaningful to study their combinations. In this paper, we introduced a combination of systematic polar codes and probabilistic shaping, and proposed that NSPC will eliminate the influence of probabilistic shaping. At the same time, the shaping method will introduce redundancy in the source part, so that the performance of NSPC for shaping is poor, while systematic polar codes can remain the shaping gain brought by probabilistic shaping. The application of NSPC and shaped systematic polar codes to multi-level coded modulation systems can be further study.

References

1. Arikan, E.: Channel polarization: a method for constructing capacity-achieving codes for symmetric binary-input memoryless channels. IEEE Trans. Inf. Theory **55**(7), 3051–3073 (2009)
2. Arikan, E.: Systematic polar coding. IEEE Commun. Lett. **15**(8), 860–862 (2011)
3. İşcan, O., Böhnke, R., Xu, W.: Probabilistic shaping using 5G new radio polar codes. IEEE Access **7**(99), 22579–22587 (2019)
4. Iqbal, S., et al.: Probabilistically shaped rate-adaptive polar-coded 256-QAM WDM optical transmission system. J. Lightwave Technol. **38** (7), 1800–1808 (2020)
5. Zhou, L., He, H., Zhang, Y., et al.: Enhancement of spectral efficiency and power budget in WDN-PON employing LDPC-coded probabilistic shaping PAM8. IEEE Access **8**(99), 45766–45773 (2020)
6. Guo, T.C., Zhang, Z., Zhong, C., et al.: A low complexity encoding algorithm for systematic polar codes. IEEE Commun. Lett. **20**(7), 1277–1280 (2016)
7. Tal, I., Vardy, A.: List decoding of polar codes. IEEE Trans. Inf. Theory **61**(5), 2213–2226 (2015)
8. Chen, Y.F., He, H.L., Zhao, X., Cao, Z.Q., Zhou, L., Dong, Z.: Low-complexity probabilistic shaping based on bit-weighted distribution matching in DMT-WDM-PON. Optics Express **28**(15), 21814–21824 (2020)
9. Ma, X., Ping, L.: Coded modulation using superimposed binary codes. IEEE Trans. Inf. Theory **50**(12), 3331–3343 (2004)
10. Sarkis, G.P., Giard, A., Vardy. C., Thibeault. W., Gross, J.: Fast polar decoders: algorithm and implementation. IEEE J. Selected Areas Commun. **32**(5), 946–957 (2014)

Turbo Equalization Technique for Data Link Communication Systems

Ao-Shuang Yang, Shi-Ming Li, Yi-Jie Wu, Yue-Qi Wang, and Ju-Hao Tan

Abstract The development of data link has greatly changed the operation mode of modern war. The complexity of wireless communication channel makes the high rate of data transmission become one of the important research topics in data link communication technology. In order to enhance the reliability of the system, channel equalization technology can be used to eliminate ISI (Inter Symbol Interference). Equalization is similar to "whitening filter", which is to "smooth" the frequency response of the channel. Different from the traditional equalization method, the equalization and decoding is combined in the Turbo equalization technology. By recycling the soft information, the bit error rate can be reduced. The Turbo equalization technology can make full use of known information. The soft information can be recycled in each iteration. The simulation results show that the performance of data link communication system can be improved by the Turbo equalization technique. This paper mainly studies the Turbo equalization technology for data link communication systems.

1 Introduction

Due to the influence of multipath transmission and fading, inter symbol crosstalk ISI often occurs in the channel. Channel equalization technology is a method to deal with this characteristic of the channel.

In the mid-1960s, adaptive filtering technology was introduced into equalization technology, and zero forcing algorithm was proposed under the peak distortion criterion [1]. In the early 1980s, Ungerboeck proposed the minimum mean square error algorithm [2], the algorithm has the advantages of small computation and good stability, but the disadvantage is slow convergence. In recent years, turbo code technology has gradually matured. Inspired by the idea of Turbo code, Douillard et al. Proposed the idea of Turbo Iterative Equalization [3, 4]. Bauch et al. proved the good effect of Turbo Equalization Technology Based on map algorithm [5],

A.-S. Yang (✉) · S.-M. Li · Y.-J. Wu · Y.-Q. Wang · J.-H. Tan
Shanghai Aerospace Electronic Technology Institute, Shanghai 201100, China

L. C. Jain et al. (eds.), *Advances in Wireless Communications and Applications*,
Smart Innovation, Systems and Technologies 299,
https://doi.org/10.1007/978-981-19-2255-8_10

but its huge amount of computation is the obstacle that it cannot be well applied in practice. T. Oberg et al. Proposed using linear equalizer instead of decision feedback equalizer to avoid the error transmission problem of Turbo equalizer [6].

The work is arranged as follows: Sect. 2 gives the principle of Turbo equalization. The algorithm of Turbo equalizer is introduced in Sect. 3. In Sect. 4, the system model is presented, and according to the results of simulation analysis are carried out.

2 Turbo Equalization Theory

2.1 Transmitter Principle

Turbo equalization technology is based on the idea of Turbo code encoding and decoding. Therefore, the transmitter principle of Turbo equalization is similar to the Turbo code encoding part, which is mainly composed of encoder, interleaver and modulator modules (Fig. 1).

Firstly, the data a_k with length K enters the encoder for encoding, and the encoder outputs the data b_k with length N. Then, it enters the interleaver. The main purpose of interleaving is to reduce burst errors. The main principle is to rearrange the information data and change the order. The output result of the interleaver c_k is converted into a sequence x_k for transmission after appropriate modulation.

2.2 Receiver Principle

The receiving part of Turbo equalization is similar to the decoding part of Turbo code and is not done all at once. The receive port is mainly composed of equalizer and decoder. It is connected by interleaver, deinterleaver and modem to transfer soft information to each other.Also the receiving part form a multi-iteration process to complete the final equalization and decode.

As shown in Fig. 2, in the first equalization process, there is no prior information. So the equalizer only calculates the value Y_k output through the channel to obtain the posterior information of the transmitted symbol $L_E^{pos}(\hat{x}_k)$, and the posterior information gets the soft information $L_D^{pri}(\hat{x}_k)$ after the deinterleaver module. Then the output information of the decoder is deducted from the input information of the decoder, and the prior information needed in the iterative process $L_E^{pri}(\hat{x}_k)$ is obtained after the interleaver. Then the prior information and the output value of the channel enter the equalizer together, and the first iteration process begins. After several iterations, no new external information is generated, and the iteration process can be stopped. Finally, the output of the decoder $L_D^{pos}(\hat{x}_k)$ should be decided, as shown in the formula:

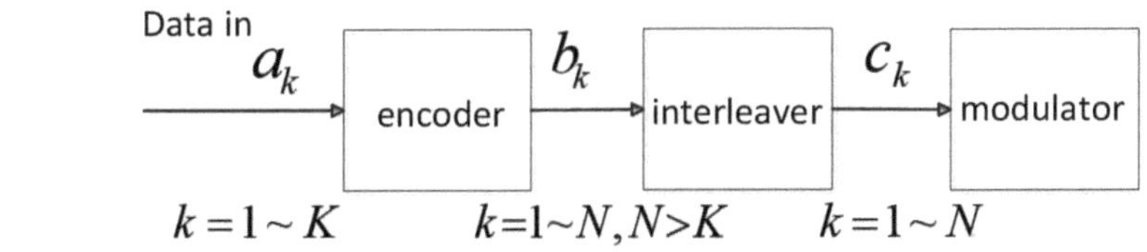

Fig. 1 Transmitter structure

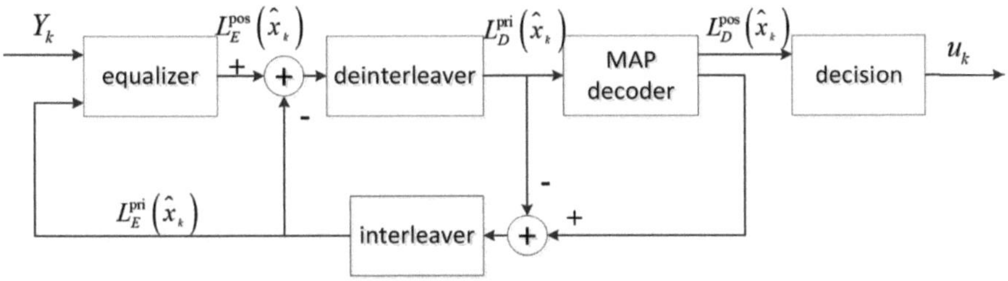

Fig. 2 Receiver structure

$$u_k = \begin{cases} 1, L_D^{\text{pos}}(\hat{u}_k) \geq 0 \\ 0, L_D^{\text{pos}}(\hat{u}_k) < 0 \end{cases} \tag{1}$$

In the whole process of receiving part, each module transmits. soft information to each other, that is, the method of SISO (Soft Input Soft Output) equalization decoding [3]. It is expressed in the form of logarithmic likelihood ratio, which avoids the error transmission due to the new interference caused by direct transmission of information.

3 Turbo Equalization Algorithm

3.1 Maximum a Posterior Probability (MAP) Algorithm

If the impulse response length of the channel is L, according to Watterson model [7], it can be equivalent to a discrete delay linear circuit with a delay of $L-1$.That is to say, the state of the channel is 2^{L-1}. The classical Prokais B channel is taken as an example for analysis, and its delay circuit structure is shown in Fig. 3. BPSK (Binary Phase Shift Keying) modulation is adopted. Without the influence of noise, the output v_k at time k channel is:

$$v_k = \sum_{l=0}^{L} h_l x_{k-l} \ , \ k = 1, 2, \cdots N \tag{2}$$

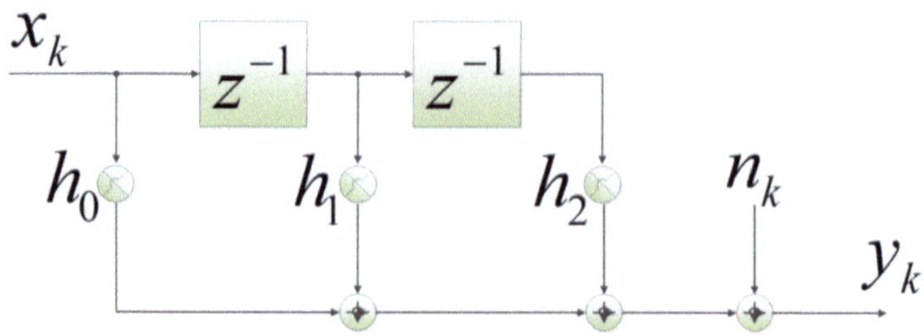

Fig. 3 Equivalent delay circuit structure of Prokais B Channel

where x_k is input signal, h_l is taps of channel.

Therefore, the output of the moment channel is determined by the channel taps, and the state transfer path is presented in a grid diagram (see Fig. 4). The four states of the channel are $s_0 = (1, 1)$ $s_1 = (-1, 1)$ $s_2 = (1, -1)$ $s_3 = (-1, -1)$. These four states can be expressed as sets $S = \{s_0, s_1, s_2, s_3\}$.

Based on the input and output values in the grid diagram and the transition path of the state, the desired posterior probability value $P(x_k|\mathbf{y})$ can be calculated.

Then, after processing by equalizer, the soft information of the sequence, namely LLR (Log Likelihood Ratio) value $L(c_k|\mathbf{y})$, is obtained:

$$L(c_k|\mathbf{y}) = \ln \frac{\sum_{\forall(s_i,s_j)\Rightarrow x_{i,j}=+1} \alpha_k(s_i) \cdot \gamma_k(s_i, s_j)\beta_{k+1}(s_j)}{\sum_{\forall(s_i,s_j)\Rightarrow x_{i,j}=-1} \alpha_k(s_i) \cdot \gamma_k(s_i, s_j)\beta_{k+1}(s_j)} \tag{3}$$

Among them, the posterior probability is calculated according to the joint distribution of the current state and the next time state. To make $\alpha_k(S_k) = p(S_k, y_1, y_2, \cdots y_{k-1})$, $\gamma_k(S_k, S_{k+1}) = p(S_{k+1}, y_k|S_k)$, $\beta_{k+1}(S_{k+1}) = p(y_{k+1}, y_{k+2}, \cdots y_N|S_{k+1})$. And $\alpha_k(S_k)$ $\beta_k(S_k)$ can be obtained by recursive calculation as follows:

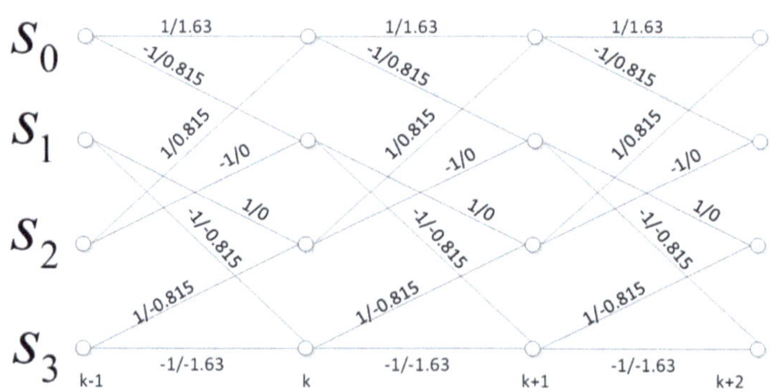

Fig. 4 Arrow diagram of Prokais B Channel

$$\alpha_{k+1}(S_{k+1}) = \sum_{\forall S_k \in S} \alpha_k(S_k)\gamma_k(S_k, S_{k+1}) \tag{4}$$

$$\beta_k(S_k) = \sum_{\forall S_{k+1} \in S} \beta_{k+1}(S_{k+1})\gamma_k(S_k, S_{k+1}) \tag{5}$$

$$\gamma_k\left(s_i, s_j\right) = \begin{cases} P\left(x_k = x_{i,j}\right) \cdot p\left(y_k \middle| v_k = v_{i,j}\right), & (i, j) \in \beta \\ 0, & (i, j) \notin \beta \end{cases} \tag{6}$$

As the calculation method is based on all the observed objects, the performance of MAP algorithm is relatively the best among the equalization algorithms. However, the above analysis shows that the shortcomings of MAP algorithm are also obvious, and it is difficult to realize due to the large amount of calculation in practical engineering practice.

3.2 Minimum Mean Square Error (MMSE) Algorithm

The Turbo equalization based on MMSE algorithm reduces the space of observed samples compared with MAP algorithm, so the calculation of equalization can be reduced to a certain extent. Turbo equalization based on MMSE algorithm is SISO algorithm. Instead of the traditional MMSE equalization method, it adopts an iterative method. Each iteration recalculates the soft information to be transmitted, and then changes the filter coefficient value in real time.

As mentioned above, the MAP algorithm-based Turbo equalization calculates the LLR value of the output posterior probability according to the observation sample space composed of all observed values. The size of the new observation space is determined by the order of linear MMSE filter. After the observation sample space is reduced, the calculation method of LLR value of posterior probability in MMSE algorithm is shown in formula (7):

$$L_E^{pos}(x_k) = \ln \frac{\Pr\left(x_k = +1 \middle| \hat{x}_k\right)}{\Pr\left(x_k = -1 \middle| \hat{x}_k\right)} \tag{7}$$

where $\hat{x}_k$ is the estimated value of x_k under MMSE algorithm. Since the receiver is not aware of the data sent by the sender x_k, the sent data is estimated. If the mathematical expectation $\overline{x}_k$ and variance v_k at the time k are known, the estimated value $\hat{x}_k$ of the sent data can be calculated:

$$\hat{x}_k = \overline{x}_k + v_k s^H \left(\sigma_w^2 I_N + H V_k H^H\right)^{-1} (z_k - H\overline{x}_k) \tag{8}$$

The tap coefficients of the channel are expressed by matrix $\boldsymbol{H}$, and the mathematical expectation $\overline{\boldsymbol{x}}_k$ and variance $\boldsymbol{V}_k$ are expressed by vector.

$$\boldsymbol{H} = \begin{bmatrix} h_{M-1}h_{M-2}\cdots h_0 0\cdots 0 \\ h_{M-1}h_{M-2}\cdots h_0 \\ \ddots \ddots \ddots \\ 0\cdots 0 h_{M-1}h_{M-2}\cdots h_0 \end{bmatrix} \tag{9}$$

$$\overline{\boldsymbol{x}}_{\boldsymbol{k}} = \left[\overline{x}_{k-M-N_2+1}, \cdots, \overline{x}_{k+N_1}\right]^T \tag{10}$$

$$\boldsymbol{V}_{\boldsymbol{k}} = Diag\left(v_{k-M-N_2+1}, \cdots, v_{k+N_1}\right) \tag{11}$$

To make $\boldsymbol{s} = \boldsymbol{H}\left[\boldsymbol{0}_{1\times(N_2+M-1)}, \boldsymbol{1}, \boldsymbol{0}_{1\times N_1}\right]^{\boldsymbol{T}}$.

The prior information of the current moment is set to 0, which meets the requirements of Turbo principle. The LLR calculation of external information no longer carries its own information, that is, the estimated value $\hat{x}_k$ and $L_D^{pri}(x_k)$ mutual independence of the transmitted information. $\hat{x}_k$ Is changed into:

$$\hat{x}_k = \boldsymbol{s}^{\boldsymbol{H}}\mathrm{Cov}(\boldsymbol{z_k}, \boldsymbol{z_k})^{-1}(\boldsymbol{z_k} - \boldsymbol{H}\overline{\boldsymbol{x}}_{\boldsymbol{k}} + (\overline{\boldsymbol{x}}_{\boldsymbol{k}} - \boldsymbol{0})\boldsymbol{s}) \tag{12}$$

The LLR of the external output of the equalizer is:

$$L_E^{ext}(x_k) = \frac{2\boldsymbol{c}_k^{\mathbf{H}}(z_k - \boldsymbol{H}\overline{\boldsymbol{x}}_k + \overline{\boldsymbol{x}}_k\boldsymbol{s})}{\left(1 - \boldsymbol{s}^{\boldsymbol{H}}\boldsymbol{c}_k\right)} \tag{13}$$

One of them $\boldsymbol{c}_k = \left[\sigma_w^2\boldsymbol{I}_N + \boldsymbol{H}\boldsymbol{V}_k\boldsymbol{H}^{\boldsymbol{H}} + (1 - v_k)\boldsymbol{s}\boldsymbol{s}^{\boldsymbol{H}}\right]^{-1}\boldsymbol{s}$.

The Turbo equalization based on MMSE algorithm reduces the observation sample space and results in less computation than MAP algorithm, but at the same time, it also pays the price of inferior performance to MAP algorithm. In order to ensure the accuracy of decoding, the channel decoding module based on MMSE algorithm Turbo equalization still adopts MAP algorithm.

4 Performance Simulation Analysis

4.1 Simulation Environment and Parameter Setting

The encoder is (7, 5) convolution code encoder, the interleaver is 1024 bit cyclic interleaver, and the constellation mapping adopts QPSK (Quad Phase Shift Keying) modulation mode. In consideration of the effect of multipath, this paper adopts ISI channel smodel with Gaussian white noise, and carries out simulation under different SNR (Signal Noise Ratio). The SISO-MAP decoder is used for channel decoding.

4.2 *Simulation Analysis:*

Simulation performance of different equalization methods

In this paper, the traditional ZF (Zero Force) equalization, LMS (Least Mean Square) equalization [8] and Turbo equalization using iterative theory are simulated respectively. As shown in Fig. 5, the effects of these equalization methods are bad in low SNR. As the SNR increases, the effects of the traditional equalization methods are still poor, and the BER (Bit Error Rate) decreases slowly with a size of about 10^{-1}, which basically has no effect on improving system performance. However when the SNR increases, BER can be effectively reduced by the Turbo equalization method, especially when the iterative curve drops faster.

Simulation performance of Turbo equalization algorithms

Figure 6 shows the BER curves of Turbo equalization using MAP algorithm and LMMSE algorithm respectively. It can be seen from the figure that the Turbo equalization based on MAP algorithm has better performance. However the MAP equalization algorithm computational complexity with the increase of the number of iterations to exponential growth, because the algorithm complexity is too high, on the practical engineering implementation requires great amount of calculation, a waste of time, and Turbo equalization algorithm based on LMMSE (Linear Minimum Mean Squared Error) rules, can greatly reduce the computational complexity, performance is also better than other algorithms, so comprehensive analysis of the algorithm is easier to project implementation.

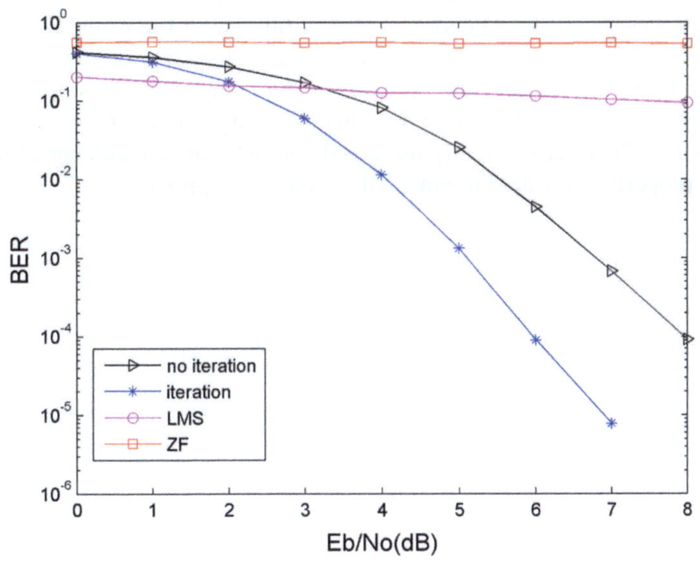

Fig. 5 BER performance comparison in different equalization methods

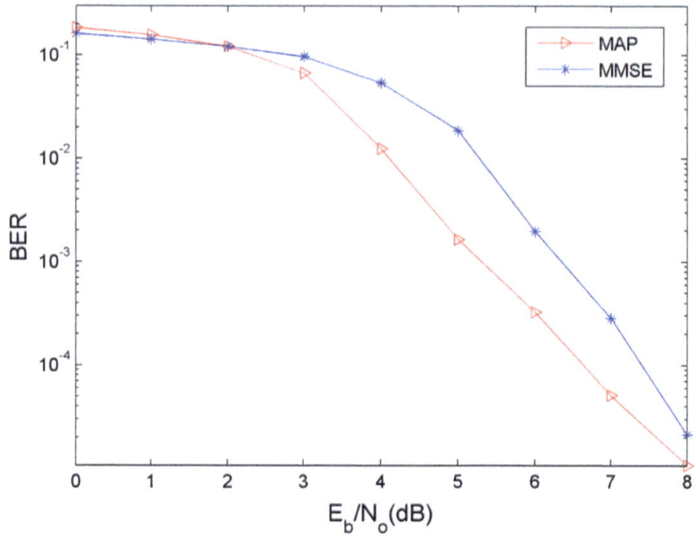

Fig. 6 BER performance comparison of different Turbo equalization algorithms

Turbo equalization simulation performance of different iterations

The number of iterations of Turbo equalizer is set to 4 at the receiving end, and the BER curve is shown in Fig. 7. According to the characteristics of BER curve, it is found that the curve can be divided into three parts: low performance loss, waterfall area and error flat layer area. With the increase of iteration times, the system performance improves gradually. The system makes full use of the external information and gradually reduces the bit error rate of the system by adopting joint equalization and decoding. However, with the increase of iteration times, the improvement of system performance is no longer obvious. Considering that the increase of iteration times also pays the price of greatly increased computation, the number of iterations should be properly controlled in practical engineering applications.

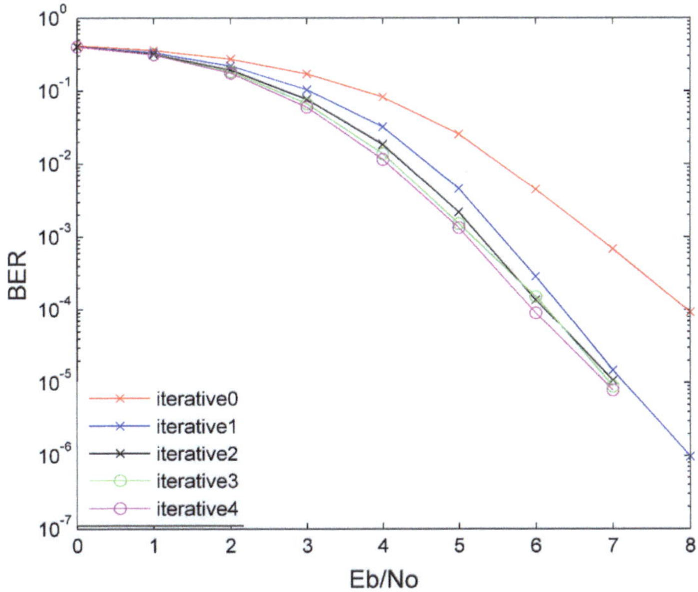

Fig. 7 BER performance comparison of different iterations

5 Conclusion

In this paper, Turbo equalization algorithm of data link communication system and multi-path effect of wireless communication channel are analyzed. Considering the engineering realization, LMMSE equalization algorithm with less computation is combined with soft input and soft output decoder to simulate the system. The simulation results show that the Turbo equalization method improves the system performance greatly when the number of iterations increases within 4 iterations.

References

1. Wibrow, B., Steinbuch, K.: A critical comparison of two kinds of adaptive classification networks. IEEE Trans. Electron. Comput. **14**(5), 737–740 (1965)
2. Frrara, E.: Fast implementations of LMS adaptive filters. IEEE Trans. Acoust. Speech Signal Process. **28**(4), 474–475 (1980)
3. Douillard, C., Jezequel, M., Berrou, C., et al.: Iterative correction of intersymbol interference: turbo equalization. Eur. Trans. Telecommun. **6**(5), 507–511 (1995)
4. Tüchler, M., Koetter, R., Singer, A.C.: Turbo equalization: principles and new results. IEEE Trans. Commun. 754–767 (2002)
5. Bauch, G., Khorram, H., Hageneuer, J.: Iterative equalization and decoding in mobile communications system. In: Second Europen Personal Mobile Communications Conference(EPMCC'97), pp. 307–312 (1997)

6. Oberg, T., Nilsson, B., Olofsson, N., et al.: Underwater communication link with iterative equalization. OCEANS 2006. IEEE (2006)
7. Gong, Z.T., Chen, G.X., Ren, X.L., Cao, J.S.: Image retrieval method based on convolutional neural network and hash coding. J. Intell. **11**(3), 10 (2016)
8. Watterson, C.C., Juroshek, J.R., Bensema, W.D.: Experimental confirmation of an HF channel model. IEEE Trans. Commun. Technol. **18**(6), 792–803 (1970)

Design of Laser Holographic Digital Image Compensation Resource Cloud Storage Platform

Li Bixiang, Wang Shuang, and Wei Yunzhu

Abstract Aiming at the problem of image quality affected by laser holography distortion caused by laser diffraction, a cloud storage platform for laser holography compensation resource is designed. Users are sent by the data access layer database access engine laser holographic digital image compensation resources storage after the request to the application of the interface layer, application interface layer in information management, platform, operation and monitoring in the interface to select the corresponding interface to cloud storage software layer, the layer through the HDFS distributed database storage technology to complete compensation resources storage, access, management and response of the laser holographic digital image compensation request, including compensation through image compensation module is complete. All data will be sent to the hardware facilities layer, which will send user demand data to the user interface and save it in the storage server to realize data cloud storage. Experimental results show that the platform can effectively compensate the laser holographic digital image, and the image quality after compensation is high, and the storage time is only 511 ms when the storage capacity is 3000 M, with high storage performance.

L. Bixiang (✉) · W. Yunzhu
Department of Information Engineering, Wuhan Institute of City, Wuhan 430083, China
e-mail: lbx153@163.com

W. Shuang
School of Computer Science, Wuhan Qingchuan University, Wuhan 430204, China

L. Bixiang
School of Computer Science and Artificial Intelligence, Wuhan University of Technology, Wuhan 430070, China

L. C. Jain et al. (eds.), *Advances in Wireless Communications and Applications*,
Smart Innovation, Systems and Technologies 299,
https://doi.org/10.1007/978-981-19-2255-8_11

1 Introduction

Holographic digital storage technology is a large-capacity storage technology born with the development of science and technology. This technology uses a photosensitive medium as a storage carrier to realize data storage based on laser interference technology [1]. With the continuous development and improvement of holographic digital storage technology, holographic digital storage technology has been applied in many different fields. Holographic digital storage technology is a technology that combines computer technology, spatial light modulator technology, and image processing technology [2]. Images obtained by laser holographic digital technology can display a large amount of image information. Therefore, through laser holographic digital technology The acquired images have special visual effects [3].

Laser holographic digital image (CCD) is an entity image amplified by laser. When observing the entity through a laser microscope, the image will be distorted due to laser diffraction, which affects the quality of laser holographic digital image [4]. Therefore, an effective laser holographic digital image is designed. Compensation for the resource cloud storage platform is of great significance.

Cloud storage is a new mode of storing data through the Internet. It is a new concept based on the development of cloud computing. Cloud storage integrates various technologies such as distributed processing, grid computing and parallel processing. A new storage technology. The development of cloud storage technology has changed the traditional network storage model. With the continuous development of informatization in my country and the popularization of the Internet, traditional storage methods can no longer meet the ever-expanding amount of information and data. With the increase in the amount of information transmission [5], people pay more and more attention to the security of information transmission. As a storage method with high security, the storage method has gradually become popular in human life. Design a cloud storage platform for laser holographic digital image compensation resources to solve the shortcomings of traditional laser holographic digital image compensation resource storage platforms such as poor security, small capacity and lack of interaction ability, improve the quality of laser holographic digital images and have high real-time data storage and data storage query function.

2 Cloud Storage Platform Design

2.1 Cloud Storage Platform Architecture

The overall architecture of the laser holographic digital image compensation resource cloud storage platform is shown in Fig. 1.

As can be seen from Fig. 1, cloud storage management is located at the central node of the overall architecture, and the central node and resource nodes are in a cluster relationship. The platform's resource storage services are provided through

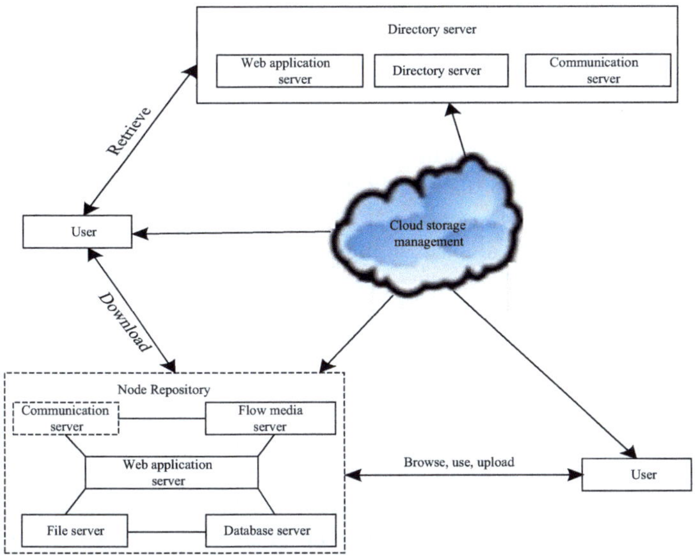

Fig. 1 Overall architecture of cloud storage platform

the node resource library [6], and resource information is sent to the central node. Nodes implement cloud storage management. The laser holographic digital image compensation resource cloud storage platform mainly uses HDFS technology as the main technology for the operation of the cloud storage platform.

Through the application interface layer, the functions of information management, platform operation and monitoring of various businesses in the cloud storage platform are realized. The storage, access, management of laser holographic digital image compensation resources and laser holographic digital image compensation are realized through the cloud storage software layer [7], and the cloud storage software layer is realized by HDFS distributed database storage technology. Hardware facilities such as storage servers and network storage are managed through the hardware facility layer.

After the user sends a resource storage request in the database access engine of the data access layer, the data access layer calls the corresponding interface through the application interface layer and transmits the request to the cloud storage software layer, and the cloud storage software layer completes the relevant instructions according to the user's requirements [8], and Send the required data to the hardware facility layer, send the final result to the user interface through the hardware facility layer, and save the cloud storage of the realization data in the storage server. The digital resource public service retrieval technology is selected and applied to the application interface layer [9] to improve the accuracy of the cloud storage platform to find information in massive information.

2.2 *Cloud Storage Software Layer Implementation*

2.2.1 HDFS Distributed Database Storage Technology

The cloud storage software layer of the system selects the distributed architecture of Hadoop and uses the HDFS distributed database storage technology to realize the storage and management of laser holographic digital images.

The HDFS distributed database storage technology has the advantages of high fault tolerance and matching with various hardware devices. The HDFS distributed database is used to store the unprocessed laser holographic image data set in the cloud storage platform.

The HDFS distributed database storage technology is a storage technology based on the Hadoop architecture. It manages resource data through the Hadoop architecture, and uses the resource data to perform file system naming, customer read and write requests, data block creation, deletion, replication and other commands [10].

Read the uncompensated laser holographic digital image data from the HDFS distributed database, and divide it into small data sets, process the data of each small data set in parallel through the laser holographic digital image compensation technology [11], and store it after all processing is completed. to a distributed data platform.

The laser holographic image compensation resource cloud storage platform uses the network to realize the compensation, storage and access services of laser holographic digital image resources. Therefore, an encryption system is set up in the HDFS distributed database to ensure the security of data storage and access [12].

Let the input attribute set of the platform be ξ, set the random number λ_j, satisfy each attribute of the laser holographic digital image $i \in \xi$, form the key through the random number λ_j, group and decrypt the platform data through the key:

$$\xi = g^{\frac{\alpha+\lambda}{\beta}}, \forall j \in \xi \tag{1}$$

In formula (1), g represents the generator of the prime number group, and the decryption process of the data belongs to the recursive process, so the intermediate function formula is as follows:

$$\begin{aligned} \eta_x &= \frac{E(D_i, C_x)}{E\left(D_i', C_x'\right)} \\ &= \frac{e\left(g^{\lambda} \cdot H(i)^{\lambda}, g^{q_x(0)}\right)}{e\left(g^{\lambda_i} \cdot H(i)^{\lambda}, g^{q_x(0)}\right)} = e(g)^{\lambda q} \end{aligned} \tag{2}$$

In formula (2), E represents encryption operation, D represents small decryption operation, and C represents ciphertext, $H(i)$ represents the hash function, which $i \in \xi$ means that the decryption packet is realized at that time, and $i \notin \xi$ means that the decryption packet fails at that time.

2.2.2 Laser Holographic Digital Image Compensation

The laser holographic digital image is an enlarged real image. The deviation caused by the distortion of the laser holographic digital image is compensated by the image compensation module [13], and a clear and high-quality laser holographic digital image is obtained. The cloud storage software layer image compensation module in the cloud storage platform compensates The process is as follows:

$$\phi_{MO}(m,n) = \exp\left[\frac{i\pi}{\lambda d_1}\left(m^2\Delta x_i^2 + n^2\Delta y_i^2\right)\right] \tag{3}$$

In formula (3), x represents the abscissa, y represents the ordinate, m and n represent the parameters corresponding to the abscissa and ordinate respectively, $d_1 = d_i - f$, d_i represents the distance between the laser digital holographic image and the laser microscope, and f represents the focal length of the laser microscope.

By $\frac{1}{f} = \frac{1}{d_0} + \frac{1}{d_i}$ expressing the relationship between the object and the image, it can be known that the quadratic phase factor formula in the laser holographic digital image is as follows:

$$\phi_{MO}(m,n) = \exp\left[\frac{i\pi}{\lambda C}\left(m^2\Delta x_i^2 + n^2\Delta y_i^2\right)\right] \tag{4}$$

Combining formula (3), formula (4) and the relationship between objects, we can get:

$$\frac{1}{C} = \frac{1}{d_i}\left(1 + \frac{d_0}{d_i}\right) \tag{5}$$

It can be seen that the conjugate term of the quadratic phase factor of formula (3) needs to be placed before the reproduction of the digital holographic image [14], so as to compensate the phase distortion caused by the laser microscope. Using the digital phase mask to represent the conjugate term, we can get:

$$\phi(m,n) = \exp\left[-\frac{i\pi}{\lambda C}\left(m^2\Delta x_i^2 + n^2\Delta y_i^2\right)\right] \tag{6}$$

According to formula (6), it can be known that the phase distortion can be compensated by selecting the optimal parameter d_0,d_1, that is, the optimal parameter C.

In the holographic digital storage device, the distance between the laser holographic digital image and the laser microscope is fixed, and the value obtained from the object-image relationship is also fixed. The above analysis shows that the distortion fringes in the phase can be eliminated by adjustment, and the laser holographic digital image compensation can be realized. [14].

Holographic digital storage devices usually realize holographic image reading through reference light. The reference light R_D formula is as follows:

$$R_D(k,l) = A\exp\left[i\frac{2\pi}{\lambda}\left(k_x k\Delta x + k_y l\Delta y\right)\right] \tag{7}$$

In formula (7), k_x, k_y both represent the vector component in the reference light. In the process of reading the laser holographic digital image, the reference light needs to be realized in the Fresnel integration [15], so the reproduced laser holographic digital image changes when the value changes, which affects the compensation phase distortion results. It can be seen that in order to ensure the quality of phase distortion compensation, the reference light needs to be moved outside the Fresnel integral. The specific process is as follows:

The formula for the properties of laser modulation is as follows:

$$\begin{aligned}&\zeta_\tau\left[\exp(i2\pi vx)f(x)\right] = \exp(i2\pi vx_i)\times\\&\exp\left(-i\pi v^2\tau^2\right)\zeta_\tau[f(x)]\left(x_i - v_x\tau^2, y - v_y\tau^2\right)\end{aligned} \tag{8}$$

In formula (8), τ and f and represent the two-dimensional parameters of the Fresnel transform and the frequency of the light wave respectively; are the adjustment coefficients of the length and width of the laser holographic digital image.

The wavefront reproduction formula is:

$$\begin{aligned}&\psi(x_i, y_i) = -i\phi(x_i, y_i)\cdot\exp(i2\pi d/\lambda)\\&\cdot R'(x_i, y_i)\zeta_\tau[I](x_i, y_i)\end{aligned} \tag{9}$$

In formula (9),$\psi(x_i, y_i)$ and $\phi(x_i, y_i)$ respectively represent the wavefront reproduction phase and the phase of the holographic digital image; and respectively represent the reference light correction coefficient and the recorded holographic digital image.

The discretization formula (10) can be obtained:

$$\begin{aligned}&\psi(m,n) = AR'(k,l)\phi(m,n)\exp\left[\frac{i\pi}{\lambda d}\left(m^2\Delta x_i^2 + n^2\Delta y_i^2\right)\right]\times\\&FFT\left\{I(k,l)\exp\left[\frac{i\pi}{\lambda d}\left(m^2\Delta x^2 + l^2\Delta y^2\right)\right]\right\}\end{aligned} \tag{10}$$

In formula (10),d and λ represent the reproduction distance and wavelength of the holographic image, respectively, Δx and Δy represent the sampling interval of the holographic plane, and represent the phase mask and the corrected reference light, respectively. The corrected reference light formula is as follows:

$$R'(k,l){=}R_D(k,l)\cdot\exp\left\{-\frac{i\pi\lambda}{d}\left[\left(\frac{k_x}{\lambda}\right)^2 + \left(\frac{k_y}{\lambda}\right)^2\right]\right\} \tag{11}$$

3 Experimental Results and Analysis

In order to detect the laser holographic digital image compensation resource cloud storage platform designed in this paper to compensate the laser holographic digital image and the storage validity, the software of seafile v6.2.11.0 is used to build the platform of this paper, and the platform of this paper is used to compensate 5 laser holographic digital images. The detection platform compensates for the validity of the laser holographic digital image.

Five initial laser holographic digital images were selected in the experiment, as shown in Fig. 2.

The digital image of laser holography after compensation using the platform in this paper is shown in Fig. 3.

Comparing Figs. 2 and 3, it can be seen that the original laser holographic digital image has poor image quality due to the light diffraction generated by the laser microscope, while the five laser holographic digital images compensated by the platform in this paper have bright colors, clear boundaries, and the image quality is improved is larger, which verifies the compensation effect of the platform in this paper.

The time used to compensate 5 images is shown in Table 1.

From Table 1, it can be seen that the time used to compensate the five laser holographic digital images using the platform in this paper is all less than 900 ms, which shows that the platform can quickly and effectively compensate the laser holographic digital images, and has high real-time performance.

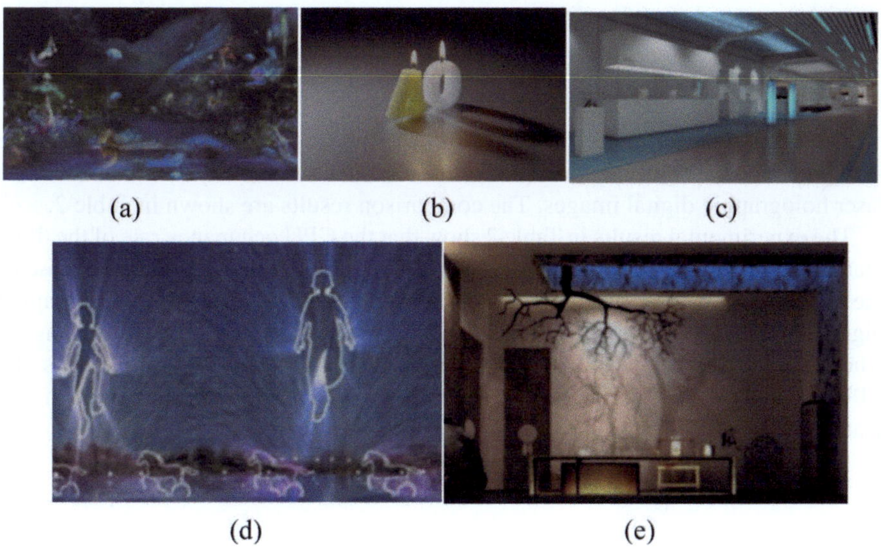

Fig. 2 Original laser holographic digital image

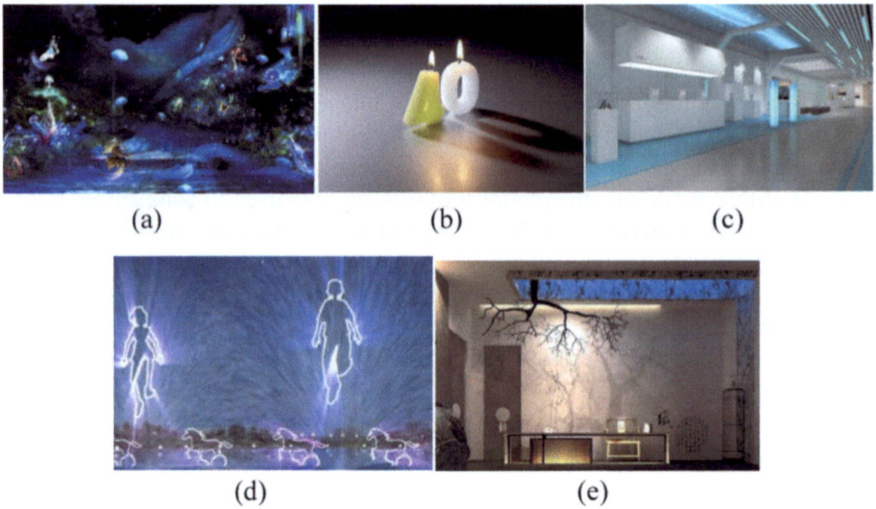

Fig. 3 Original laser holographic digital image

Table 1 Compensation time for 5 laser holographic digital images

Image name	Encryption time/ms	Decryption time/ms	Compensation time/ms
A	105	214	853
B	123	205	796
C	115	186	847
D	136	235	736
E	141	307	804

The storage performance of the platform in this paper is further tested, and the storage performance of the three platforms is set to continuously store 1000 100 kb laser holographic digital images. The comparison results are shown in Table 2.

The experimental results in Table 2 show that the CPU occupancy rate of the three platforms for storing laser holographic digital images increases with the increase of the number of images, but the CPU occupancy rate for storing laser holographic digital images using this platform is the lowest for different numbers of images., when using this platform to store 1000 100 kb laser holographic digital images, the CPU occupancy rate is only 2.33%, which verifies the storage performance of this platform.

Table 2 Comparison of CPU share of different platforms

Laser holographic digital image quantity/piece	This article platform%	Big data platform%	Dynamic encryption platform%
100	1.25	1.83	1.76
200	1.42	2.05	2.11
300	1.53	2.33	2.45
400	1.61	2.41	2.67
500	1.58	2.62	2.81
600	1.71	2.73	2.93
700	1.81	2.96	3.17
800	1.93	3.15	3.35
900	2.01	3.25	3.47
1000	2.33	3.65	3.72

4 Conclusion

In order to effectively improve the poor image quality of laser holographic digital images due to laser diffraction, a cloud storage platform for laser holographic digital image compensation resources is designed, the HDFS distributed database storage technology is applied to the cloud storage platform, and a large number of experiments are carried out to verify the performance of the platform. Effectiveness, the experimental results show that the use of this platform can not only effectively compensate the laser holographic digital image, but also make the compensated laser holographic digital image high in definition, bright in color, large in storage space, high in efficiency, and less time-consuming. The storage platform has great advantages.

Acknowledgements This work was supported by seven fund projects: 2020 Natural Science Foundation of Hubei Province (Research on User Behavior Network Circle Phenomenon and Public Opinion Guidance Mechanism, No: 2020CFB795); 2021 Hubei Provincial Scientific Research Project (Research on the application of face recognition technology in mobile terminal, No: B2021421); 2021Wuhan Institute of City Scientific Research Project (Design of mobile phone check-in system for face recognition, No: 2021CYYBKY003); the second batch of industry-university-research cooperation in 2021 Collaborative education project (Python programming and advanced data analysis practice, No: 202102235013); the first batch of industry-university-research cooperation collaborative education projects in 2020 (Research on the network talent training scheme based on the "New H3C" experimental platform, No: 202002034014, Teachers' artificial intelligence application ability improvement plan under the background of "Internet", No: 202002295002); 2021 Wuhan Institute of City key teaching and research projects: (Research on practical teaching of information security cloud platform under the background of "Internet+", No: 2021CYZDJY008); 2021 Wuhan Institute of City general teaching and research project number: (Research on the quality control system of blended teaching course construction in application-oriented colleges and universities, No: 2021CYYBJY005).

References

1. Chao, G., Yongfu, W., Haobo, C., et al.: Automatic phase distortion compensation algorithm in digital holography. J. Opt. **38**(12), 105–111 (2019)
2. Zi, L.A., Xiaoying, R., Zhang, et al.: Imaging through scattering medium based on speckle illumination and holography. J. Opt. **37**(8), 135–142 (2017)
3. Lihan, G., Xinke, W., Yan, Z.: Terahertz digital holographic imaging of biological tissues. Opt. Precis. Eng. **25**(3), 611–615 (2017)
4. Wen, X., Yang, L., Feng, P., et al.: Automatic phase aberration compensation method for digital holography microscopy combined with scribe fitting and deep learning. Acta Photonica Sinica **47**(12), 164–173 (2018)
5. Yun, P., Weiqing, P.: Design and application of Michelson interferometer based on digital holography technology. Appl. Opt. **39**(1), 93–99 (2018)
6. Yimin, G., Liujie, S.: Adaptive holographic watermarking algorithm combining Retinex and HVS. Opt. Technol. **43**(6), 555–560 (2017)
7. Liao Shuhong, W., Jing, Z.H., et al.: Research on real-time pre-distortion correction technology for binocular digital images. Electro-Opt. Control **25**(5), 113–118 (2018)
8. Liang, L., Ningfang, S., Di, F., et al.: Coupling loss analysis of polarization-maintaining fiber and Y-waveguide based on digital image. China Laser **45**(11), 225–231 (2018)
9. Yang Jing, W., Sijin, Z.W., et al.: Digital image correlation for full-field microstrain measurement of printed circuit boards. Infrared Laser Eng. **46**(11), 31–38 (2017)
10. Yan, Y., Gaoke, C.: Image restoration algorithm based on optical compensation and pixel-by-pixel transmittance. J. Commun. **38**(5), 48–56 (2017)
11. Yuxiang, W.L., Mingyang, M.Y., et al.: Highlight error compensation method based on high dynamic range image technology. Infrared Technol. **40**(10), 52–58 (2018)
12. Zexin, J., Yan, P.: Color compensation of underwater images based on electromagnetic theory. Adv. Lasers Optoelectron. 55(8):237–242 (2018)
13. Jing, Y., Qi, L., Wenpan, G.: Influence of control parameters of compressed sensing 3D reconstruction algorithm on reproduction of terahertz digital holography. China Laser **45**(10), 294–304 (2018)
14. Wu Kai W., Xuecheng, Z.L.: Experimental research on super-resolution digital holography. Opt. Technol. **44**(1), 101–105 (2018)
15. Shaoduo, L., Xiangdong, G., Yangjin, L., et al.: Weld deviation prediction algorithm based on neural network compensation Kalman filter. Appl. Laser **38**(06), 50–55 (2018)
16. Liangfu, L., Bin, Z., Guoliang, Z., et al.: Research on depth image inpainting and error compensation method based on optimal estimation. Appl. Opt. **39**(1), 45–50 (2018)

Research on Librarian Demand Prediction Based on the GM (1, 1) Model and BP Neural Network Combined Model

Leilei Peng, Ying Liu, and Ke Chen

Abstract With the development of science and technology, the automation level of library is constantly improving, and the number of librarians is decreasing year by year. As one of the most important human resources to promote the sustainable development of library, librarians are particularly important, therefore, it is necessary to research its changing trend. This study constructs a GM (1, 1)-BP Neural Network combined model, and takes the library of Sichuan University as case study. The simulation results show that, compared with a single prediction model, the GM (1, 1)-BP Neural Network combined model has higher prediction accuracy and smaller errors, which can further improve the prediction accuracy.

1 Introduction

Human resources are the most important and valuable resources in promoting the sustainable development of libraries [1], and librarian is the core component of human resources. Collections, equipment, physical space, virtual space, etc. are vital to the development of the library, but it is the librarians that ultimately play a decisive role. In recent years, with the development of information technology, traditional librarians continue to learn new technologies and acquire new capabilities. Some traditional librarians are transformed into subject librarians, they integrate into the research group and teaching team to carry out subject analysis, content demonstration, achievement evaluation and so on [2]. Some traditional librarians are transformed into digital librarians to carry out data access service, data management and so on [3].

The development of information technology has brought tremendous changes to the librarian. Taking China as an example, with the continuous expansion of enrollment in universities, the pressure of various work in the library is increasing, while the number of librarians is decreasing instead of increasing. According to the

L. Peng · Y. Liu · K. Chen (✉)
Library, Sichuan University, No. 24 South Section 1, Yihuan Road, Chengdu 610065, China
e-mail: chenkechenke@scu.edu.cn

L. C. Jain et al. (eds.), *Advances in Wireless Communications and Applications*, Smart Innovation, Systems and Technologies 299,
https://doi.org/10.1007/978-981-19-2255-8_12

data released by the Steering Committee for Academic Libraries of China, in 2013, the number of librarians in 484 universities in China was 21,685, which reduced to 19,045 in 2019 [4]. Therefore, it is necessary to predict the changing trend of the librarian, so as to scientifically and reasonably allocate and regulate this key resource, and promote the development of university libraries under the background of the information era.

Based on the construction of GM (1, 1) model, BP neural network model and the GM (1, 1)-BP neural network combined model, this study conducted a case study of Sichuan University Library, and fitted the number of librarians from 2010 to 2020 and verified the combined model, and find out the law of librarian change, so as to optimize the allocation of human resources.

2 Research Method

GM (1, 1)-BP neural network combined model is constructed for research. GM (1, 1) model and BP neural network have their own advantages and disadvantages, GM (1, 1) model requires a small amount of data, and BP neural network is suitable for dealing with objects with complex internal structure. The use of one model only for the prediction may lead to deviations between the predicted value and the real data. Therefore, this study combines the two models for comprehensive prediction.

Based on the prediction of GM (1, 1) model and BP neural network, this study obtains the final weight by determining the weight of the single prediction model. According to the scholar's previous research [5], residual sum of squares (RSS) can be used to reflect the accuracy of the model. The smaller the RSS, the higher the accuracy of the model. Therefore, we determine the weight of the combined model by minimizing the RSS.

For w_i is the weight of the ith prediction, $i = 1, 2 \ldots j$. In this study, we use two prediction models, the first is GM (1, 1) model, and the second is BP neural network. The error corresponding to the combined sequence at time t is δ_t, the corresponding error of the ith prediction at time t is $\delta_{i(t)}$. Q is the RSS of combined prediction errors.

The steps of constructing combined model are:

$$\delta_t = L_{(t)} - x_{(t)} \tag{1}$$

where, $L_{(t)}$ is the predicted value, and $x_{(t)}$ is the real data.

Constrained objective function Q:

$$\text{Min}\, Q = \sum_{t=1}^{n} \left|\delta_t^2\right| = \sum_{t=1}^{n} \left|\sum_{i=1}^{j} \left(w_i \delta_{i(t)}\right)^2\right| \tag{2}$$

Subject to:

$$\sum_{i=1}^{j} w_i = 1, 0 \leq w_i \leq 1 \tag{3}$$

The combined model is:

$$y_t = w_1 L_1(t) + w_2 L_2(t) + \cdots + w_j L_j(t) \tag{4}$$

3 Empirical Study

3.1 Brief Introduction of Research Case

This research takes the Sichuan University Library as the case study. Sichuan University Library is the library with the longest history and the largest scale of documents in Southwest China, it is composed of Arts and Science Library, Engineering Library, Medical Library and Jiang'an Library. The library has a total of 8.17 million paper documents, 312 electronic document databases, 3.68 million electronic books, 120,000 electronic journals, and 190,000 h of audio and video. In 2010, the library of Sichuan University collected 5.94 million documents, which increased to 8.17 million in 2020, and the annual expenses increased from about 17.95 million in 2010 to about 43.01 million in 2020. The substantial increase in the collection of documents and expenses will inevitably lead to an increase in the work content, but the number of librarians has dropped from 225 to 162 in the past 11 years. Therefore, Sichuan University Library should allocate human resources reasonably and explore a scientific human resources management system.

3.2 Data Source and Standardization

This study selects the data of Sichuan University Library from 2010 to 2020 as the sample data, data source is the statistical data of Steering Committee for Academic Libraries of China. The explanatory variables are the purchase cost of literature resources (x_1), the purchase cost of electronic resources (x_2), the purchase cost of paper resources (x_3), and the annual expenses (x_4), and the explained variable is the number of librarians (y). The explanatory variables are closely related to the explained variable, the higher the costs, the greater the workload of the library, the more librarians are needed. The purchase cost of literature resources includes the purchase cost of electronic resources, paper resources, multimedia materials, etc.

Table 1 Data standardization

Year	y	x_1	x_2	x_3	x_4
2010	1.0000	0.0000	0.1509	0.0000	0.0000
2011	0.9048	0.0946	0.2511	0.0195	0.0372
2012	0.8413	0.0222	0.0986	0.1828	0.0234
2013	0.6667	0.1622	0.2515	0.2429	0.1131
2014	0.5873	0.6834	0.7922	0.3357	0.7684
2015	0.4762	0.9060	0.9220	0.7527	0.8478
2016	0.3968	0.1130	0.0000	0.7970	0.1173
2017	0.3810	1.0000	0.9179	1.0000	0.8731
2018	0.1905	0.8266	1.0000	0.2498	0.7423
2019	0.1111	0.9172	0.9618	0.6786	1.0000
2020	0.0000	0.7708	0.9209	0.3032	0.7598

The annual expenses include document purchase cost, document binding and repair resource cost, equipment asset purchase cost, equipment and facility maintenance cost, office cost, etc.

The dimensions of each index are different and cannot be directly compared, so it needs to be dimensionless, and the range transformation method is used to make the index dimensionless, as shown in formula 5:

$$y_{ij} = \frac{x_{ij} - \min\limits_{1 \le i \le n}\{x_{ij}\}}{\max\limits_{1 \le i \le n}\{x_{ij}\} - \min\limits_{1 \le i \le n}\{x_{ij}\}} \tag{5}$$

where, y_{ij} is the value of the j evaluation index of the i sample which has been dimensionless, x_{ij} is the original index value of the i index of the j sample.

3.3 Results

3.3.1 Predicting with GM (1, 1) Model

According to the GM (1, 1) model prediction steps [6], the explanatory variable is substituted into MATLAB 2020a version to calculate the prediction results, as shown in Fig. 1.

The prediction results of GM (1, 1) model are shown in Fig. 1. The average relative error of GM (1, 1) prediction value is 1.6143e−4. In this study, the posteriori error method was used to test the suitability of the model, S1 = 431.7636, S2 = 4.2618, and the posterior error ratio C = 0.0099, $P > 0.9$, according to the accuracy level, the prediction accuracy of the GM (1, 1) model is excellent.

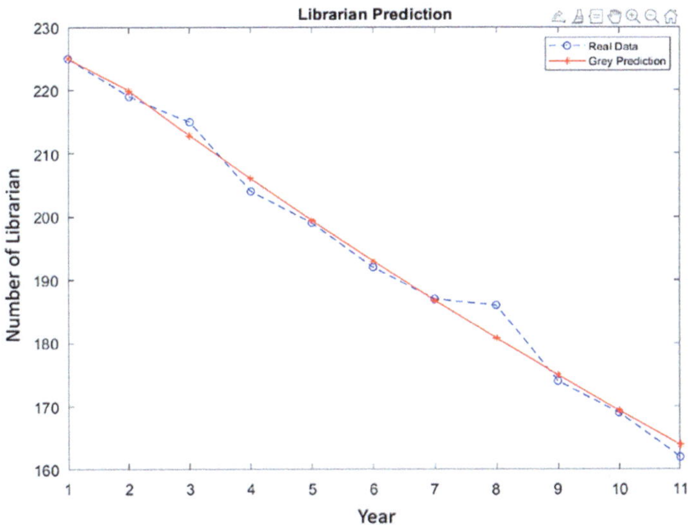

Fig. 1 Number of librarians prediction: GM (1, 1) model

3.3.2 Predicting with BP Neural Network

We import the standardized data into MATLAB 2020a version for neural network training, the number of input layers is 4, the number of hidden layers is 4, and the number of output layer is 1. In this study, sigmoid function is used to calculate the hidden layer and MATLAB purelin function is used to calculate the output layer, and the maximum number of iterations is set to 1000, the target accuracy is set to 1×10^{-4}, training rate is set to 0.1, and the Levenberg–Marquardt (trainlm) is used to training algorithm. This algorithm has faster convergence speed and higher training accuracy than other algorithms. The ratio of training samples to test samples is 3:1, that is, the number of librarians from 2010 to 2017 is the training samples, and the number of librarians from 2018 to 2020 is the test samples. The training results are shown in Fig. 2.

The BP neural network training performance is illustrated in Fig. 2, which shows that when the learning times of BP neural network is 28, the curve begins to converge, and the prediction accuracy is 6.9572e−04, which meets the preset accuracy requirements.

BP neural network prediction results are shown in Fig. 3, the result shows that the difference between the real data and the predicted value is small, indicating that the training effect is good and the prediction result is reasonable.

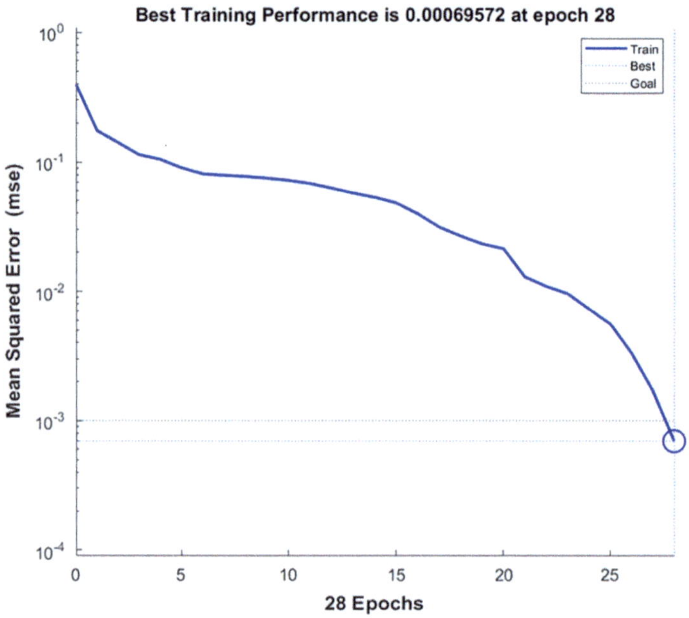

Fig. 2 BP neural network training performance

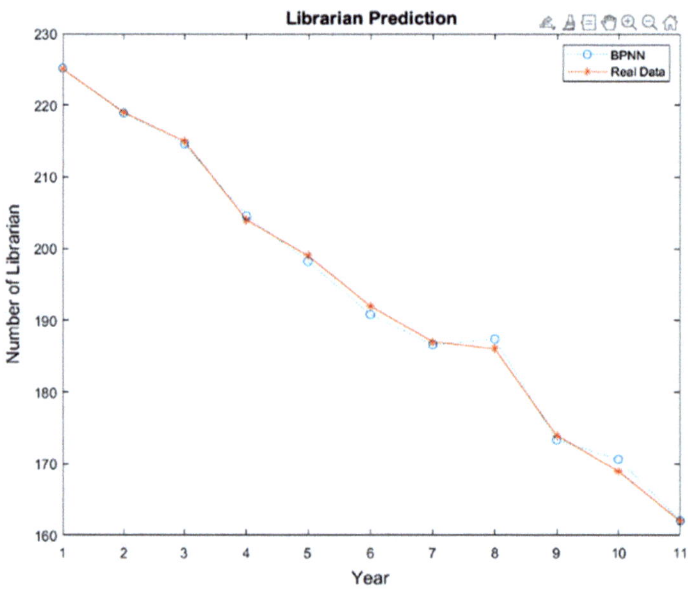

Fig. 3 Number of librarians prediction: BP neural network

Table 2 GM (1, 1)-BP neural network weight table

Model	GM (1, 1)	BPNN
Weight	0.140350877	0.859649123

3.3.3 Predicting with GM (1, 1)-BP Neural Network

According to the construction method of GM (1, 1)-BP neural network combined model, the weight of GM (1, 1)-BP neural network model is calculated, as shown in Table 2.

According to the weight value of GM (1, 1) model and BP neural network in Table 2, a combined prediction model is constructed:

$$y_t = 0.140350877L_{1(t)} + 0.859649123L_{2(t)} \tag{6}$$

We take the data predicted by GM (1, 1) model and BP neural network into formula 6 to calculate the results of the combined model, on this basis, compared with the predicted value calculated by GM (1, 1) model and BP neural network model alone, the results are shown in Table 3.

Table 3 shows that since 2010, the number of librarians of Sichuan University Library has continued to decrease, from 225 in 2010 to 162 in 2020, with a decrease rate of 28%. It can be predicted that the number of librarians will continue to decline in the next few years. The average relative error of GM (1, 1) model fitting value and BP neural network fitting value is 0.71% and 0.38%, respectively. It shows that the prediction accuracy of the two models is high. By analyzing the relative error of a single prediction model, it can be found that GM (1, 1) model has low prediction

Table 3 Comparison of prediction

Year	Real data	Predicting value			Relative error		
		GM (1, 1)	BPNN	GM (1, 1)-BPNN	GM (1, 1) (%)	BPNN (%)	GM (1, 1)-BPNN (%)
2010	225	225	225	225	0.00	0.00	0.0
2011	219	220	219	219	0.46	0.00	−0.06
2012	215	213	214	214	−0.93	−0.47	−0.53
2013	204	206	205	205	0.98	0.49	0.56
2014	199	199	198	198	0.00	−0.50	−0.43
2015	192	193	191	191	0.52	−0.52	−0.37
2016	187	186	186	186	−0.53	−0.53	−0.53
2017	186	180	187	186	−3.23	0.54	0.01
2018	174	175	173	173	0.57	−0.57	−0.41
2019	169	169	170	170	0.00	0.59	0.51
2020	162	163	162	162	0.62	0.00	0.09

accuracy for abnormally fluctuating data, and the prediction accuracy of BP neural network has always remained stable. The average relative error of the fitting value of the combined model is 0.32%, which further improves the prediction accuracy of the model and can be better used to predict the change trend of the librarian.

4 Discussion

In this study, the prediction accuracy of GM (1, 1) model is excellent, according to Yao's research, the growth rate of a sequence is generally constant, while a monotonically decreasing sequence is susceptible to random factors, which will lead to undesirable changes. Therefore, when the data sample is small, the sequence changes are more stable [6]. In this work, the data is a monotonically decreasing small sample data, which can maintain relative stability, so the prediction accuracy is high. However, among the three models, the relative error of GM (1, 1) model is still greater than that of BP neural network and the combined model, and the relative error in 2017 was the largest, which was 3.32%. The possible reason is that GM (1, 1) model is more suitable for linear data, but the change of the number of librarians is affected by many factors. It is a non-linear system, so the prediction accuracy is lower than the other two models. The prediction relative error of the BP neural network is in the range of $[-1\%, 1\%]$, and the prediction accuracy is generally higher than that of the GM (1, 1) model. The purchase cost of literature resources, electronic resources, paper resources and annual expenses included in this study directly affect the workload of the library. The higher the four indicators, the greater the workload, the more librarians are needed. On the contrary, the lower the workload, the less librarians are needed. Similar to Li's research [7], we found that the combined model of GM (1, 1)-BP neural network is better than the single GM (1, 1) model and BP neural network, and the fitting performance is better. Some scholars' research shows that both GM (1, 1) model and BP neural network have a large error when they are used separately [8], since the data in this study is a monotonically decreasing small sample data, and the explained variable is closely related to the explanatory variables, this situation did not occur.

JL Sun proposed that 5% of the value created by the library is generated by the library buildings, 20% by the information resources and 75% by the librarians [9]. The sample data of this study shows that since 2010, the Sichuan University Library has significantly increased its annual expenses and workload, while the number of librarians has been declining. At the same time, with the development of information technology, the work content and working methods of the library have undergone tremendous changes compared with the past. Therefore, librarians must constantly learn new technologies and change from traditional librarians to subject librarians and digital librarians. The library should timely organize relevant training, purchase corresponding equipment, and vigorously strengthen the enthusiasm of librarians for the application of information technology [10]. We should also note that due to the continuous influx of new technologies and knowledge, the increasing requirements for the ability of librarians, coupled with factors such as age, marriage, and readers, have brought pressure to librarians. Once this pressure continues for a long time, librarians will gradually enter a state of job burnout, which is not conducive to the development of library work [11].

There are some limitations in this study. Firstly, this study collected sample data of Sichuan University Library for only 11 years from 2010 to 2020, and the sample size is relatively small. Secondly, methods for predicting the evolution trend of time sequence data include multiple regression models, autoregressive integrated moving average model (ARIMA), etc., and this study only uses GM (1, 1) model and BP neural network. Thirdly, due to the availability of data, the explanatory variables of this study only include the purchase cost of literature resources, electronic resources, paper resources and annual expenses. Therefore, in the next step of research, scholars can use a variety of prediction methods to compare the results, and add more explanatory variables to study more cases to make the prediction results more effective.

5 Conclusion

This research constructs a GM (1, 1)-BP neural network combined model, and takes Sichuan University Library as a case study to conduct an empirical analysis. Results show that, the prediction accuracy of the GM (1, 1)-BP neural network combined model is better than that of a single prediction model. From 2010 to 2020, the number of librarians continued to decline, the library management department should pay attention to this trend and take practical measures to actively respond to the adverse effects of the reduction of librarians to promote the development of the library.

References

1. Chen, K., Jiang, X., Huang, H., et al.: An analysis of research on human resource management in university library. J. Library Sci. Soc. Sichuan **03**, 9–13 (2017)
2. Pinfield, S.: The changing role of subject librarians in academic libraries. J. Librariansh. Inf. Sci. **33**(1), 32–28 (2001)
3. Sarasvathy, P., Nambratha, G.R., Giddaiah, D.: Changing roles of the librarians in the virtual/digital era. SRELS J. Inf. Manag. **49**(5), 495–500 (2019)
4. Steering Committee for Academic Libraries of China. Statistical Data. [EB/OL]: http://www.scal.edu.cn/tjpg/tjsj. Accessed on 25 Sept 2021
5. Wu, J.L.: Study on GM(1,1)-BP combination model based on grey correlation. J Chongqing Univ Technol (Nat Sci) **33**(11), 207–210 (2019)
6. Yao, T.X., Liu, S.F., Xie, N.M.: On the properties of small sample of GM (1, 1) model. Appl. Math. Model. **33**(4), 1894–1903 (2009)
7. Li, M.H., Zhao, M.G., Liu, M., et al.: Energy consumption prediction of campus building based on the GM-BP neural network. Build Energy Eff **44**(11), 87–90 (2016)
8. Guo, D.J., Zou, Y.: Prediction of China's urban residents consumption level based on Grey GM (1, 1)-BP neural network. Adv. Serv Sci. Serv. Inf. Technol. **52**, 193–199 (2014)

9. Sun, J.L.: The importance of human resource management in library reform. Library Tribune **22**(5), 133–135 (2002)
10. Ramzan, M., Asif, M., Ahmad, S.: Librarians' attitudes towards application of information technology in academic libraries in Pakistan. Inf. Res. Int. Electron. J. **26**(1), 887 (2021)
11. Smith, D.L., Bazalar, B., Wheeler, M.: Public librarian job stressors and burnout predictors. J. Libr. Adm. **60**(4), 412–429 (2020)

Offloading Strategy of Computing Tasks in Cooperative Vehicle Infrastructure Systems

Haiying Xia, Yingji Liu, Xinlei Wei, and Guoliang Dong

Abstract In cooperative vehicle infrastructure systems (CVIS), common vehicles can offload their computing tasks that are difficult to complete in time for them to the roadside multi-access edge computing (MEC) servers for execution. However, resources of roadside MEC servers are still limited in the large-scale deployment of vehicles. To address this problem, this paper introduces the task offloading from user vehicles to surrounding resource-rich service vehicles, and constructs communication and computation models for computation offloading. Subsequently, we propose the utility function for computation offloading in which the benefits and costs of task offloading are fully considered, and then derive the optimal strategy for task offloading. The experimental results show that the latency and utility values of computation offloading with the proposed strategy are significantly better than those of other computation offloading strategies, which improves the efficiency of task execution in CVIS.

1 Introduction

In recent years, various vehicle road collaborative applications emerge in endlessly. These advanced auxiliary applications covering different aspects put forward higher requirements for vehicle computing and storage capacity [1]. However, the resources that each vehicle can use for computing and storage services are always limited, so it is not enough to support the operation of various on-board high-performance applications.. Therefore, researchers have proposed a new computing architecture of multi access edge computing (MEC) [2] to make up for the lack of resource [3]. In order to reduce the pressure on computing and storage resources of vehicles or intelligent roadside infrastructure and shorten the data end-to-end transmission delay, MEC migrates network cloud computing resources to the network edge close

H. Xia (✉) · Y. Liu · X. Wei · G. Dong
Key Laboratory of Operation Safety Technology On Transport Vehicles, Ministry of Transport, PRC, No.8, Xitucheng Road, Haidian District, Beijing 100088, China
e-mail: hy.xia@rioh.cn

L. C. Jain et al. (eds.), *Advances in Wireless Communications and Applications*,
Smart Innovation, Systems and Technologies 299,
https://doi.org/10.1007/978-981-19-2255-8_13

to various terminal devices. To make use of these computing and storage resources at the network edge, vehicles need to migrate their difficult computing tasks to the edge nodes. This process of task migration is generally called computing offload. The purpose of designing an efficient and stable edge computing task migration and offloading strategy is to make full and effective use of edge computing resources.

Although some studies have designed the communication architecture and computing offload scheme in MEC network under the vehicle road collaboration environment, most of the research methods of traditional cloud computing are simply applied to edge computing, without considering that the MEC server deployed on the roadside is also subject to many restrictions and cannot deploy too many resources, which reduces the application efficiency of the whole vehicle road collaboration system. Considering that some high-performance autonomous vehicles running on the road are also regarded as an external service resource, and the spare unused computing and storage resources on these vehicles are used to share the pressure of roadside servers, this paper proposes a computing migration and offloading destination selection strategy for ordinary vehicles in the environment where roadside MEC and on-road service vehicles can provide computing services.

2 Situation Analysis

At present, some communication architectures and computing offload schemes in MEC network under vehicle road cooperation environment have been proposed by relevant studies at home and abroad. Reference [4] studies the dynamic sharing of 5th-Generation (5G)spectrum, and designs a sharing architecture of Dedicated Short-Range Communication (DSRC) and 5G spectrum for immersive experience driven vehicle communication. Reference [5] proposes a two-level edge computing architecture for automated driving services in order to make full use of the intelligence at the wireless edge, and investigates the research challenges of wireless edge caching and vehicular content sharing and develops possible solutions. In reference [6], a big data analysis system based on MEC is proposed for the charging scenario of electric vehicles. The MEC server with mobile sensing ability interacts with the electric vehicles within its scope, so as to disseminate the predicted charging availability information of charging station, collect the big data of electric vehicle driving, and realize decentralized calculation on the basis of data mining and aggregation. In reference [7], the problem of task offloading is studied from the perspective of matching, and a price based matching algorithm is proposed to optimize the total network delay. In reference [8], deep reinforcement learning is used to design a resource allocation strategy for the joint design problem of communication, caching and computing, and the challenges of vehicle mobility and strict service deadline constraints are considered.

However, most of the studies in the above literatures simply apply the traditional research methods of cloud computing to edge computing, and seldom consider that the computing and storage resources of MEC server are not as infinite as cloud computing [1]. The MEC server deployed on the roadside is also limited by the roadside space and the deployment cost of operators, so it can not deploy too many resources. As a result, in the large-scale application environment of vehicle road collaboration, the MEC server can not serve all vehicles, which reduces the application efficiency of the whole vehicle road collaboration system. In this case, we can consider some high-performance self driving vehicles running on the road as an external service resource, and use the unused computing and storage resources on these vehicles to share the pressure of roadside servers. In this way, ordinary vehicles on the road can not only transfer their own complex computing tasks to the MEC server on the roadside, but also transfer their tasks to the high-performance service vehicles with similar trajectories around, so as to further balance the load of the whole system. Therefore, in this environment where both the roadside MEC and the surrounding service vehicles can realize the computing migration and offloading, it is necessary to further study the vehicle task migration and offloading strategy, that is, how to choose the destination of the task migration under different conditions, so as to obtain greater benefits and reduce the cost, and finally achieve the optimal state of the system.

This paper focuses on the choice strategy of the destination for the calculation migration and offloading of ordinary vehicles in the environment that both MEC and road service vehicles can provide computing services. In the process of decision-making, the vehicle will consider the benefits and costs. In terms of the benefits, the most important thing for vehicles is the final execution time after the task has been transferred and offloaded. If the task is performed faster than the vehicle locally, the vehicle can get a positive benefit. If the execution time exceeds the maximum time limit of the task, the vehicle will not benefit. The cost of vehicles is mainly measured by the price of calculation services. Because MEC on the road can provide strong computing resources, it can make the price of MEC service on the road higher than that of the service vehicles on the road in order to promote the full utilization of resources. Therefore, this paper first considers the application, communication and calculation model of edge computing and offloading which is oriented to vehicle road coordination, and takes the influence of vehicle movement into account, and then calculates the total execution time of task migration and offloading. Then, based on the vehicle task migration and offloading model, we propose the calculation migration and offloading utility function in the scenario of vehicle road collaborative edge calculation. In order to obtain the task migration and offloading strategy of each vehicle under the maximum utility of the system, we consider the task migration behavior of all vehicles. Finally, the software is used to solve the problem and the simulation results verify the advantages of the proposed method compared with other task migration and offloading schemes.

3 System Model

In the vehicle road collaborative scenario, ordinary vehicles on the road are connected to the surrounding high-performance vehicles or roadside MEC server through V2V (vehicle to vehicle) and V2R (vehicle to road) communication, and then the task migration and offloading strategy is used to select whether the task migration and offloading is to the roadside server or the surrounding vehicles that can provide services, so as to ensure the maximum utility of the whole system.

In this paper, we consider the vehicle task migration problem under the coverage of a single MEC server, and assume that the set of user vehicles in the coverage of the MEC is n = {1, 2, ..., n}. In the process of migration and offloading of vehicle computing tasks, it is necessary to determine a computing offloading model to estimate the benefits and costs. So we will propose calculation and communication models, evaluate the task execution delay when the task is transferred and offloaded to roadside MEC and surrounding service vehicles, and finally give the utility function of the system according to the revenue and cost of vehicles.

3.1 Communication Model

According to the method in reference [9], we can use three parameters to represent the application model of vehicle i which needs to calculate the migration and offloading. L_i is the size of the input data of the calculation task, α_i is the computational complexity of the task, $t_{i,max}$ represents the maximum deadline required for the task. If the task can be completed within the required maximum time limit, the vehicle will get higher revenue, otherwise it will not get revenue, and it will also have losses due to the cost. For the sake of simplification, we consider the application of each vehicle as a whole task, that is, we only consider the whole task to a single destination.

In the communication model, we consider that V2I (vehicle to infrastructure) communication mode is adopted between vehicle and roadside MEC access point, and V2V communication mode is adopted between user vehicle and service vehicle, and assume that the communication channels of these two kinds of communication are flat Rayleigh channels [9]. In this model, $d_i^{-\theta}$ represents the path loss of communication, where d_i is the distance from vehicle i to MEC access point or surrounding service vehicle, and θ is the path loss index of communication. In addition, h_i and N_0 represent channel fading factor and Gaussian white noise power respectively.

In this way, the data transmission rate between vehicle i and roadside MEC receiving point can be presented by the following formula:

$$R_{i,E} = W_i \log_2\left(1 + \frac{P_i d_{i,E}^{-\theta} |h_i|^2}{N_0}\right) \tag{1}$$

Among them, W_i is the channel bandwidth, P_i is the transmission power of the vehicle, $d_{i,E}$ indicates the distance between the vehicle and the roadside edge node when the vehicle task is transferred and offloaded.

Similarly, we can express the transfer rate of task migration and offloading between user vehicle i and its surrounding service vehicle j as follows:

$$R_{i,j} = W_i \log_2\left(1 + \frac{P_i d_{i,j}^{-\theta} |h_i|^2}{N_0}\right) \tag{2}$$

In the formula,$d_{i,j}$ represents the distance between user vehicle i and service vehicle j when the task is transferred and offloaded.

3.2 Simulation Experiment

Compared with the traditional mobile terminal devices, vehicles have more sufficient energy sources, so the vehicle road collaborative computing task migration and offloading focuses less on the energy consumption of task migration and offloading, and more on the processing delay of offloading task. In the vehicle road collaborative computing scenario, the processing delay of the offloading task mainly includes the communication transmission delay of the task and the calculation delay of the task execution. In accordance with [9], $\beta_{i,U}$ represents the overhead of the uplink in the task transmission to the roadside. Since the amount of the result data after the task execution is generally small, the magnitude of the input data can be ignored. Here, we do not consider the delay caused by the return of the task execution result. Furthermore, the uplink transmission delay of vehicle i for task migration and offloading to the roadside MEC can be expressed as:

$$t_{i,U} = \frac{\beta_{i,U} L_i}{R_{i,E}} \tag{3}$$

The execution time of the task on the roadside MEC server is:

$$\tau_{i,E} = \frac{\alpha_i L_i}{f_E} \tag{4}$$

where f_E is the computing power of MEC server. Therefore, the total processing delay of task migration and offloading from the user's vehicle i to the roadside MEC server is as follows:

$$t_{i,E} = \frac{\alpha_i L_i}{f_E} + \frac{\beta_{i,U} L_i}{W_i \log_2\left(1 + \frac{P_i d_{i,E}^{-\theta} |h_i|^2}{N_0}\right)} \tag{5}$$

Similarly, the total processing delay of task migration and offloading from user vehicle i to its surrounding service vehicle j is as follows:

$$t_{i,V} = \frac{\alpha_i L_i}{f_j} + \frac{\beta_{i,req} L_i}{W_i \log_2\left(1 + \frac{P_i d_{i,j}^{-\theta} |h_i|^2}{N_0}\right)} \quad (6)$$

$\beta_{i,req}$ represents the cost of the task to offload the request to the service vehicle j. In this way, we can get the time when the vehicle i's computing task is transferred to MEC server or peripheral service vehicle j based on the computing offload model. Then, we take these data results to build the utility function of vehicle computing task migration and offloading, in which we consider the time benefit of task execution and the price cost of computing service:

$$u_i(a_i) = \frac{t_{i,max} - t_{i,V}}{t_{i,max}}(1 - a_i) + \frac{t_{i,max} - t_{i,E}}{t_{i,max}} a_i - (1 - a_i)\rho_j p_E - a_i p_E \quad (7)$$

Among them, a_i is the final calculation migration behavior of the vehicle, if $a_i = 1$, the vehicle will move to the roadside MEC. $a_i = 0$ indicates that the vehicle is migrating to the surrounding vehicles, p_E is the price of task migration and offloading to the roadside MEC for calculation, ρ_j is the ratio of the price calculated by the task in the surrounding vehicles and the price calculated by the MEC on the roadside. Therefore, in order to ensure the load balance of the system, the optimal migration and offloading strategy pursued by all ordinary vehicles can be expressed as follows:

$$\boldsymbol{a}^* = \arg\max_{\boldsymbol{a}} \sum_{i=1}^{N} u_i(a_i) \quad (8)$$

In this way, vehicles can make decisions according to this optimal task migration strategy, so as to balance the computing pressure of roadside MEC server in large-scale vehicle environment and maximize the global utility.

4 Analysis of Experimental Results

In this section, we use series of simulation results to verify the performance of the proposed strategy. Specifically, we use MATLAB to solve the optimal strategy in formula (8), and compare the performance of the strategy with other schemes in terms of average task delay and utility value. The relevant experimental parameters are shown in Table 1.

Firstly, we compare the average utility values of different task migration and offloading schemes through several experiments, and the results are shown in Fig. 1. Using the optimal strategy proposed in this paper, the average utility value of vehicles

Table 1 Experimental parameters

Parameters	Implication	Value
L	The input data size of the task	1 Mbits
α	Task calculation complexity	240 cycles/bit
f_j	Computing power of peripheral service vehicles	2 GHz
f_E	Roadside MEC computing capacity	5 GHz
θ	Path fading factor	2
W	Communication bandwidth	10 MHz
P	Device transmit power	0.2 W
$\beta_{i,U}$	Overhead of uplink in vehicle and roadside transmission	1
$\beta_{i,req}$	The overhead of task offloading request to service vehicle j	1
ρ_j	The price ratio of vehicles transferring and offloading tasks to peripheral service vehicles and roadside MEC servers	0.7

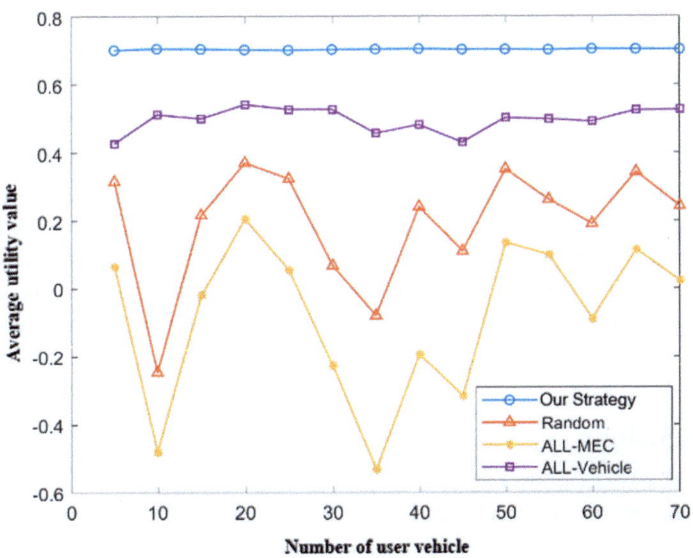

Fig. 1 Variation of average utility value with the number of user vehicle in different strategies

can be maintained at a high level. However, the average utility value of the random strategy and the other two fixed migration and offloading strategies is obviously low, and due to the influence of vehicle movement and channel fading, the utility value is constantly fluctuating and extremely unstable.

Figure 2 shows the average time delay of task migration and offload under different migration and offload schemes. In this experiment, the maximum migration and offloading time required by the user's vehicle $t_{i,max}$ is set to 0.6 s. Through comparison, we can see that the average delay of the optimal strategy proposed in this paper

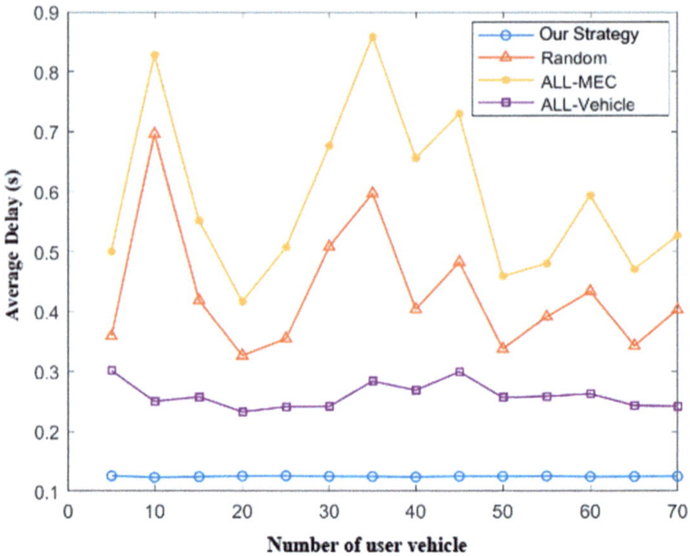

Fig. 2 The change of expected delay with the number of users' vehicles in different strategies

will not exceed the maximum time required by the task under different number of users' vehicles, and can continue to be stable at a lower value. At the same time, due to the continuous changes of vehicle movement and channel state, the average delay can not be maintained at a stable level by the random strategy and the other two fixed migration and offloading strategies. Therefore, this experiment shows the advantage of the task migration and offloading strategy in controlling the task execution delay.

5 Conclusions

In this paper, the task migration and offloading behavior of each vehicle to the roadside MEC in the vehicle road collaborative scenario is regarded as the occupation of server resources. In order to ensure that the roadside MEC server will not exceed the load in the large-scale vehicle deployment environment, the vehicle workshop computing task migration and offloading is introduced to share the pressure of the roadside server. Furthermore, this paper presents the application, communication and computing model of vehicle computing task migration and offloading in vehicle road collaborative scenario, and proposes the utility function of vehicle task migration and offloading decision, and deduces the optimal strategy. According to the revenue and cost of task migration estimated by this function, vehicles can decide to migrate tasks to roadside MEC or peripheral service vehicles. Through series of experiments, we compare the average utility and time delay of this strategy with other task migration and offload schemes, and verify the performance improvement of the proposed strategy. In the future, the model will be extended to multi vehicle and multi MEC collaboration environment to maximize its utility.

Acknowledgements This paper is supported by the Central Public Research Institutes Special Basic Research Foundation (2021-9065b).

References

1. Wang, Y., Lang, P., Tian, D., et al.: A game-based computation offloading method in vehicular multiaccess edge computing networks. IEEE Internet Things J. **7**(6), 4987–4996 (2020)
2. Taleb, T., Samdanis, K., Mada, B., et al.: On multi-access edge computing: a survey of the emerging 5G network edge cloud architecture and orchestration. IEEE Commun. Surv. Tutorials **19**(3), 1657–1681 (2017)
3. Pham, X.Q., Nguyen, T.D., Nguyen, V.D., et al.: Joint node selection and resource allocation for task offloading in scalable vehicle-assisted multi-access edge computing. Symmetry **11**(1), 58 (2019)
4. Zhou, H., Xu, W., Bi, Y., et al.: Toward 5G spectrum sharing for immersive-experience-driven vehicular communications. IEEE Wirel. Commun. **24**(6), 30–37 (2017)
5. Yuan, Q., Zhou, H., Li, J., et al.: Toward efficient content delivery for automated driving services: an edge computing solution. IEEE Netw. **32**(1), 80–86 (2018)
6. Cao, Y., Song, H., Kaiwartya, O., et al.: Mobile edge computing for big-data-enabled electric vehicle charging. IEEE Commun. Mag. **56**(3), 150–156 (2018)
7. Liu, P., Li, J., Sun, Z.: Matching-based task offloading for vehicular edge computing. IEEE Access **7**, 27628–27640 (2019)
8. Hu, R.Q.: Mobility-aware edge caching and computing in vehicle networks: a deep reinforcement learning. IEEE Trans. Veh. Technol. **67**(11), 10190–10203 (2018)
9. Wang, Y., Sheng, M., Wang, X., et al.: Mobile-edge computing: partial computation offloading using dynamic voltage scaling. IEEE Trans. Commun. **64**(10), 4268–4282 (2016)

Analysis and Implementation of Safe Transmission and Cluster Expansion Based on Tomcat

Guoliang Dong, Haiying Xia, and Fujia Liu

Abstract With the widespread application of Internet technology, higher requirements are put forward for the security of websites. The traditional plaintext data transmission method will lead to information leakage, and there are great security risks. This article takes the security deployment of an in-use website as an example, analyzes the basic deployment architecture of the website, and analyzes the organization of the tomcat architecture. On the basis of HTTP deployment, the deployment of secure transmission is implemented, which improves the security of data interaction. On the basis of single-instance deployment, the deployment of Tomcat cluster expansion has improved the performance of website operation. After the actual operation of the website, the above optimization strategy has been operating stably.

1 Introduction

With the development of new technologies, efficient and convenient Internet website construction methods are applied to different website application scenarios. The security of the transmission information and the high efficiency of the transmission process are still the basic requirements of network information transmission. HTTP Protocol is a simple request-response protocol. Based on the client/server model, it specifies what kind of message the client may send to the server and what kind of response it gets [1, 2]. Since HTTP uses a clear text transmission method and does not check the integrity of the message, an attacker can intercept the user's important data using interception tools. With the development of network security technology, the HTTPS (Hyper Text Transfer Protocol over Secure Socket Layer) technology capable of secure transmission has become increasingly mature and has been rapidly applied in different scenarios. HTTPS is composed of HTTP and SSL (Transport Layer Security)/TLS (Transport Layer Security), and information encryption is performed on the basis of HTTP transmission. After information transmission is encrypted by

G. Dong (✉) · H. Xia · F. Liu
Research Institute of Highway Ministry of Transport, Beijing 100088, China
e-mail: gl.dong@rioh.cn

L. C. Jain et al. (eds.), *Advances in Wireless Communications and Applications*,
Smart Innovation, Systems and Technologies 299,
https://doi.org/10.1007/978-981-19-2255-8_14

TLS, the security of the transmission is improved and the shortcomings of HTTP transmission are made up [3].

When a website is developed and deployed on the Internet, it is generally necessary to use an HTTP server to exchange information between the application and the user. The combination mode of Apache HTTP Server (Apache)/Nginx and Apache Tomcat (Tomcat) is the current mainstream website application deployment method. Apache or Nginx is used as a web server, mainly processing static page information of the website. Tomcat is used as an application server to receive the dynamic request information of Servlets and JSP forwarded by the Web server, and send the processed response information back to Apache, and finally Apache returns the response to the client. In order to improve the performance of the website, you can configure multiple tomcats to achieve load balancing and improve the website's response speed to user requests [4, 5].

This article takes the actual deployment of an in-use website as an example to analyze the realization and application of website secure transmission technology, and discuss the realization method of load balancing by combining Apache and Tomcat.

2 Secure Transmission and Load Balancing

2.1 Website Structure

Figure 1 shows the architecture of a conventional website. WebServer is specifically used to parse static web content such as HTML, JPG/GIF pictures, TXT, VBSCRIPT, PHP, etc. It mainly opens ports 80 and 443 to achieve HTTP and HTTPS access.

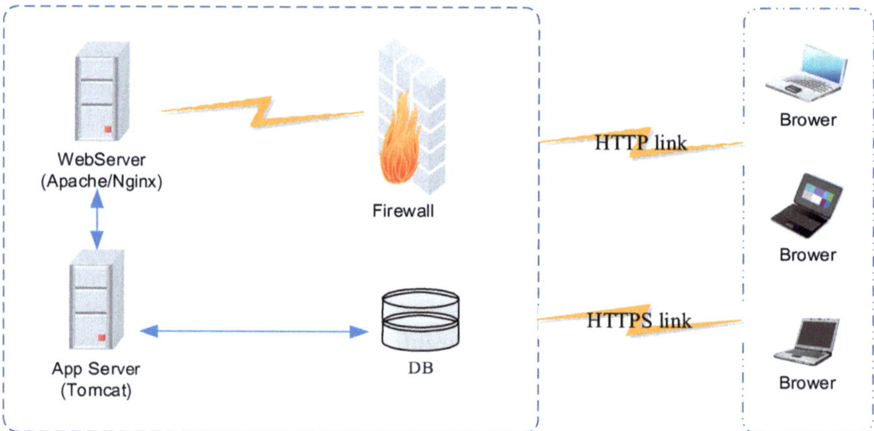

Fig. 1 Architecture diagram of a website

App Server is used to parse dynamic web pages that need to be parsed by a Java compiler, which can reduce the pressure on App Server. App Server and Web Server are connected by means of internal IP (Internet Protocol) and port, which improves the security of access [6]. App Server can also improve website performance through cluster expansion.

2.2 Website Structure

Figure 2 shows the structure of Tomcat.

As shown in Fig. 2, "Server" is the top-level container. A Tomcat instance contains only one Server, and one Server contains one or more Services. The role of Service is to assemble Connector and Container. Standard ServletRequest and ServletResponse communication are used between Connector and Container [7]. Each Service contains two parts: Connector and Container. Connector is used to process connection-related transactions and provide conversion between Socket and Request and Response. Container is used to encapsulate and manage Servlet, process Request from Connector, and return the processing result to Connector. Each Service corresponds to a Container, but can correspond to multiple Connectors. Because a Service can have multiple connections, Fig. 3 shows a schematic diagram of Tomcat connections with multiple connections and different protocols.

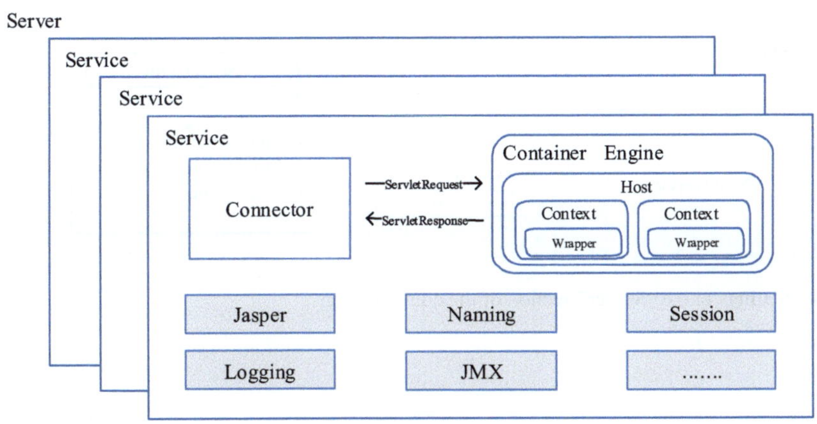

Fig. 2 Tomcat structure

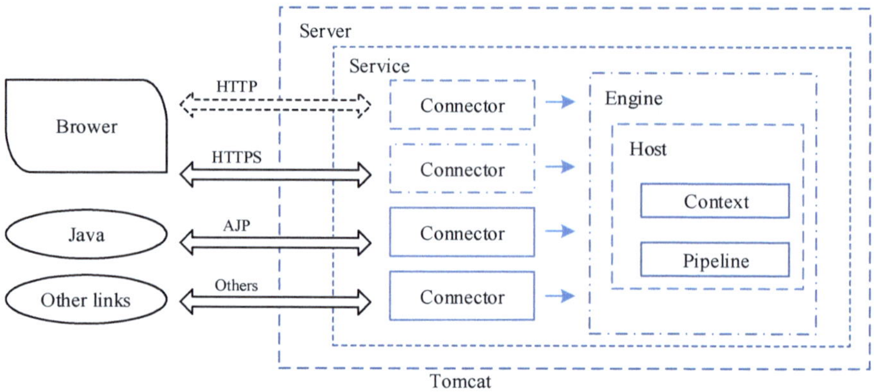

Fig. 3 Tomcat links with multiple connections and different protocols

2.3 *General Configuration of Tomcat*

The configuration file of Tomcat is "server.xml", which is generally in the "conf" folder of the Tomcat installation directory. Figure 4 shows the items that need to be configured in the "Server" part.

Figure 5 shows the items that need to be configured in the "Service" section.

The following are some general configurations in service.xml [8].

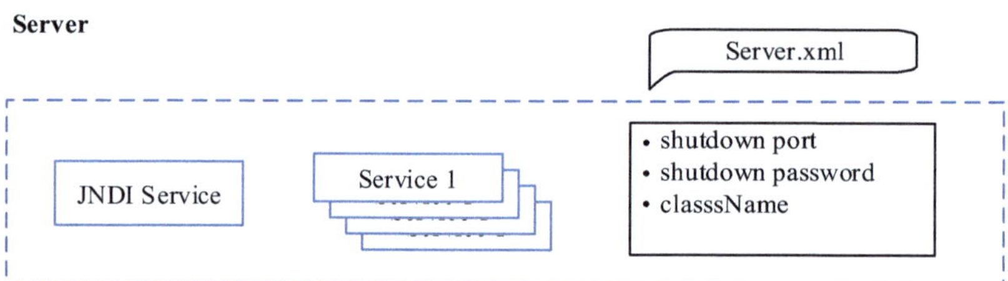

Fig. 4 Settings in the "Server" section of the configuration

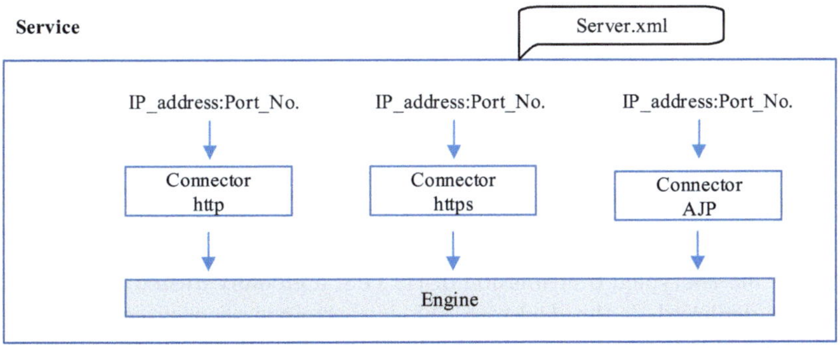

Fig. 5 Settings in the “Service” section of the configuration

```
<Server port="8005" shutdown="SHUTDOWN" debug="0">
<Connector port="8080"
    protocol="HTTP/1.1"
    minSpareThreads="6"
    maxThreads="20"
    connectionTimeout="20000"
    redirectPort="8443" />
<Connector port="7009"
    minSpareThreads="6"
    maxThreads="20"
    address="localhost"
    protocol="AJP/1.3"
    redirectPort="8443"
    secret="YOUR_AJP_SECRET" />
<Engine name="Catalina" defaultHost="localhost" jvmRoute="tom1">
<Host name="localhost"
    debug="0"
    appBase="d:\webapps"
    unpackWARs="true"
    autoDeploy="true">
```

According to the above configuration, the Tomcat Server will wait for the shutdown command at port 8005. If it receives the “SHUTDOWN” string, Tomcat closes the Service. Connector listens for HTTP requests from browsers. The minimum number of backup threads for the Connector is 6, each thread is responsible for 1 user’s request, and the maximum number of threads allowed is 20. The total number of threads does not exceed “maxThreads”. When the user request is HTTPS, the request is forwarded to port 8443. Tomcat listens to the AJP/1.3 request forwarded by Apache on port 7009, and sets the “secret” value (YOUR_AJP_SECRET) to realize the confidential transmission of information. The “Engine” part is used to process the HTTP request received by the Connector. It matches the request with the virtual host and forwards the request to the corresponding Host for processing. The default virtual host is "localhost". A Host is a virtual host, "name" is set to the domain name, and "appBase" is set to the path of the web application group.

2.4 Secure Transmission

To realize the encrypted transmission of HTTPS, a globally trusted SSL certificate should be deployed. If the deployed SSL has obtained security certification, the security lock of the user's browser will be displayed in green. There are some websites that provide free SSL certificates, but they generally have shorter periods and lower security levels [9]. You can apply for free or paid SSL according to the security level requirements of the website. Add appropriate content to the server.xml file to establish an encrypted channel to realize the secure transmission of exchanged information. "keystorePassword" is the password of the password store in the SSL certificate file "server.jks" [10].

```
<Connector port="8443"
    protocol="HTTP/1.1"
    SSLEnabled="true"
    scheme="https"
    secure="true"
    clientAuth="false"
    sslProtocol="TLS"
    keystorePass="keystorePassword"
    keystoreFile="server.jks"/>
```

2.5 Cluster Expansion

After a single Tomcat is configured and successfully debugged, only a few modifications are needed to realize the cluster expansion of Tomcat [11, 12].

Figure 6 shows a schematic diagram of cluster expansion using two tomcat instances.

For cluster expansion, Connector works in Conjunction mode, and each Tomat instance works as a "worker" of the Web server, and is identified by the port IP address and port number monitored by the instance. Tomcat runs as an independent process in an independent JVM (Java Virtual Machine), and the AJP (Apache JServ Protocol) connector receives the request from the Web server and interprets it into a format that can be processed by the Catalina engine. Figure 7 shows a schematic diagram of the connector's work.

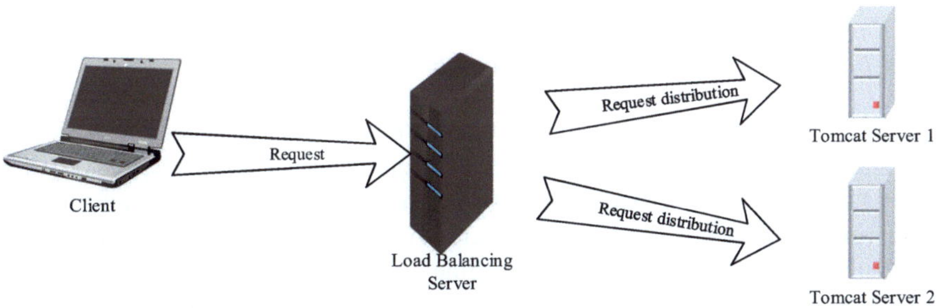

Fig. 6 Cluster expansion with 2 Tomcat instances

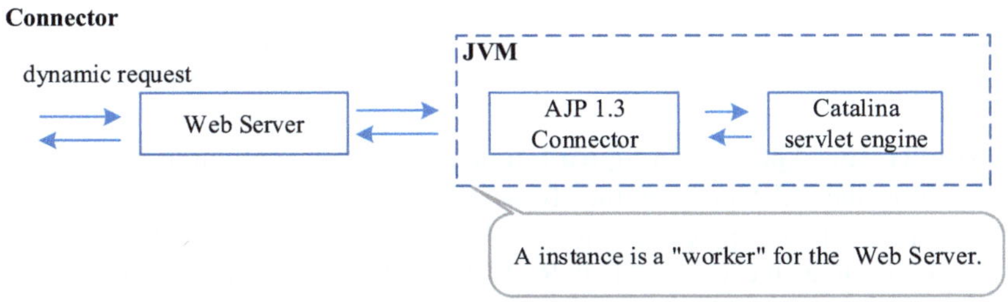

Fig. 7 Conjunction mode for the connector

Follow the steps below to achieve cluster expansion.

1. **Deploy JK in Apache Server**. Put mod_jk.so in the modules directory of Apache Server. JK is a Tomcat Connector, using JK can forward jsp/Serverlet requests from users to Tomcat for processing via Apache.
2. **Moidify Apache configure file**. Open the httpd.conf file in the Apache conf directory and add the following content to load mod_jk when Apache Server starts.
   Include conf/mod_jk.conf
3. **Configure Tomcat as a cluster extension**. The Tomcat cluster can be deployed on one or multiple machines. Table 1 is the port allocation scheme that implements the expansion of 3 Tomcat clusters.

Modify the "Engine" part of the Tomcat1 configuration file (/tom1/conf/server.xml) as follows.

Table 1 Port allocation of Tomcat cluster

Tomcat	Tomcat Server	Connector HTTP/1.1	Connector AJP/1.3
Tom1	8005	8080	7007
Tom2	8006	8081	7008
Tom3	8007	8082	7009

<Engine name="Catalina" defaultHost="localhost" jvmRoute=" tom1">
<Cluster className="org.apache.catalina.ha.tcp.SimpleTcpCluster"/>

Copy 2 copies of the first Tomcat folder "/tom1" that is successfully configured, and rename the copy folder names to "/tom2" and "/tom3" respectively, and modify "/tom2/conf/server.xml" and "/tom2/conf/server.xml" and The port number at "Server, Connector HTTP/1.1, Connector AJP/1.3" in "/tom3/conf/server.xml". Modify the value of "jvmRoute" in "Engine" to "tom2" and "tom3" respectively. Delete the content of the "Connector port = "8443" part of the "tom2" and "tom3" configuration files. The "Engine" part of the Tomcat 2 and configuration files is as follows.

<Engine name="Catalina" defaultHost="localhost" jvmRoute=" tom2">
<Engine name="Catalina" defaultHost="localhost" jvmRoute=" tom2">

4. **Modify the Apache configuration file**. Add the configuration content to monitor the AJP port in the Tomcat cluster. Modify the content in the Apache/conf/work.properties file as follows, where the value of "secret" should be the same as the value ("YOUR_AJP_SECRET") of the "secret" setting of "Connector AJP/1.3" in the Tomcat server.xml setting file.

```
worker.list = loadbalancer
worker.tom1.port=7007
worker.tom1.host=localhost
worker.tom1.type=ajp13
worker.tom1.lbfactor = 1
worker.tom1.secret=Your_AJP_SECRET
worker.tom2.port=7008
worker.tom2.host=localhost
worker.tom2.type=ajp13
worker.tom2.lbfactor = 1
worker.tom2.secret=Your_AJP_SECRET
worker.tom3.port=7009
worker.tom3.host=localhost
worker.tom3.type=ajp13
worker.tom3.lbfactor = 1
worker.tom3.secret=Your_AJP_SECRET
worker.loadbalancer.type=lb
worker.loadbalancer.balance_workers=tom1, tom2, tom3
worker.loadbalancer.sticky_session=1
worker.loadbalancer.sticky_session_force=0
```

3 Conclusion

This article takes the construction of an application website as an example to analyze the method of website information exchange, Tomcat's architecture and data transmission mechanism. After completing the deployment of Apache and Tomcat, the website can be run in plaintext transmission. In order to improve the security of website information transmission, the deployment plan of website security transmission is analyzed and actual deployment is carried out. Based on the analysis of the Tomcat cluster expansion operation mechanism, the configuration files of Apache and Tomcat on the website were updated, and the deployment of 3 Tomcat clusters was realized. After the above strategies are deployed on the website, the number of detected attacks has been reduced from about 2000 a month to less than 1000, and the website security protection strategy has achieved the expected effect.

Acknowledgements This paper is supported by the Central Public Research Institutes Special Basic Research Foundation (2021-9065b).

References

1. Richard, S.: TCP/IP Illustrated, Vol. 3: TCP for Transactions, HTTP, NNTP, and the UNIX Domain Protocols, 1st edn. Post & Telecom Press, Beijing (2016)
2. Xiao, J.: HTTP Packet Capture Actual Combat, 1st edn. Post & Telecom Press, Beijing (2018)
3. David, G., Brian, T., Marjorie, S., Sailu, R., Ansbu, A.: HTTP: The Definitive Guide, 1st edn. Post & Telecom Press (2012)
4. Wang, Y., Chen, W.W.: The application of Nginx instead of Apache in the high-concurrency Web load balancing system. Electron. Test **06**, 88–92 (2015)
5. Port, port mapping, Tomcat configuration file Homepage, https://blog.csdn.net/Ltoto/article/details/90040636. Last accessed 2019/05/09
6. Chopra, V., Sing, L., Genender, J.: Apache Tomcat 6. 1st edn. Post & Telecom Press, Beijing (2008)
7. Tomcat overall architecture analysis Homepage, https://www.cnblogs.com/bjguanmu/p/8874621.html. Last accessed 2018/04/18
8. Ivan, R.: The Definitive Guide to HTTPS: Deploy SSL, TLS, and PKI on Servers and Web Applications, 1st edn. Post & Telecom Press, Beijing (2016)
9. Keytool Homepage, https://docs.oracle.com/javase/6/docs/technotes/tools/windows/keytool.html
10. Qian, Z.Q.: Tomcat application server high concurrency optimization processing. Comput. Program. Skills Maintenance **000**(002), 129–136 (2018)
11. Feng, Y.Y.: Research and implementation of performance optimization of high concurrent time attendance system. Electron. Design Eng. **27**(18), 29–32 (2019)
12. Sun, R.P.: Research on the integration of Tomcat and Apache. Comput. Program. Skills Maintenance **000**(014), 6–8 (2011)

Research on Key Technologies of New Generation Vehicle Wireless Network

Xuan Dong, Bo Li, and Zhiyu Zheng

Abstract With the continuous improvement of automobile intelligence, the delay, synchronization, reliability, security and concurrency of on-board network are required to be higher. In this paper, the key technologies of the new generation of vehicle-mounted wireless network are studied. According to the characteristics of in-vehicle data communication of intelligent vehicles, the network architecture, basic characteristics of physical layer, synchronization and access, resource allocation, channel coding and modulation are deeply studied and designed.

1 Introduction

As the last link of information transmission, wireless short-range communication technology plays a huge role in all aspects of social life. Every year, more than 10 billion new wireless short-range connected devices are put into the market, accelerating the birth of a large number of new applications and value scenarios [1]. With the development of applications, the emergence of new services puts forward new requirements for the existing wireless short-range communication technology in low delay, high reliability, precise synchronization, high concurrency and information security [2]. In particular, the automotive industry is in urgent need of wireless short distance communication technology that can better match the business demand and development trend. Based on this background, this paper designs a new generation of

X. Dong (✉)
Key Laboratory of Operation SafetyTechnology on Transport Vehicles, Ministry of Transport, PRCNo. 8, Xitucheng Road, Haidian District, Beijing 100088, P.R. China
e-mail: x.dong@rioh.cn

B. Li
Huawei Technologies Co., Ltd., Huawei BLD, Bantian, Longgang, Shenzhen 518129, P.R. China

Z. Zheng
Eagle Drive Tech Shenzhen Co., Ltd., 1-B, 5 Floor, 23 Building, KeYuanXi Industrial Zone, Kejiyuan, Yuehai Street, Nanshan Area, Shenzhen 518057, P.R. China

L. C. Jain et al. (eds.), *Advances in Wireless Communications and Applications*,
Smart Innovation, Systems and Technologies 299,
https://doi.org/10.1007/978-981-19-2255-8_15

vehicle-mounted wireless network, aiming to promote the application of new wireless short-range communication technology including this technology in scenarios such as intelligent vehicles, and further promote the continuous evolution of new wireless short-range communication technology.

With the continuous improvement of automobile intelligence level, the requirements of in-car data communication are getting higher and higher [3–5]. At present, CAN (Controller Area Network) bus used by large commercial vehicles CAN not meet the current requirements in bandwidth, delay, concurrency and other aspects. Secondly, this paper studies a new generation of vehicle-mounted wireless network for the use of intelligent vehicles to meet new application requirements.

2 System Architecture

2.1 Network Architecture

Nodes in the system are classified into management nodes (G nodes) and managed nodes (T nodes). In specific application scenarios, a single G node manages a certain number of T nodes, and G nodes are connected to these T nodes to complete specific communication functions. A single G node and its connected T node together constitute a communication domain.

At this point, the CDC (Cockpit Domain Controller) and the vehicle-mounted device form a communication domain. When the mobile phone is connected to the CDC, the mobile phone can also serve as the T node in the communication domain. See Fig. 1.

In some scenarios, there may be multiple communication domains: In the environment of smart vehicles, mobile phones can also be used as G nodes to connect to wearable devices. At this point, mobile phones and wearable devices form

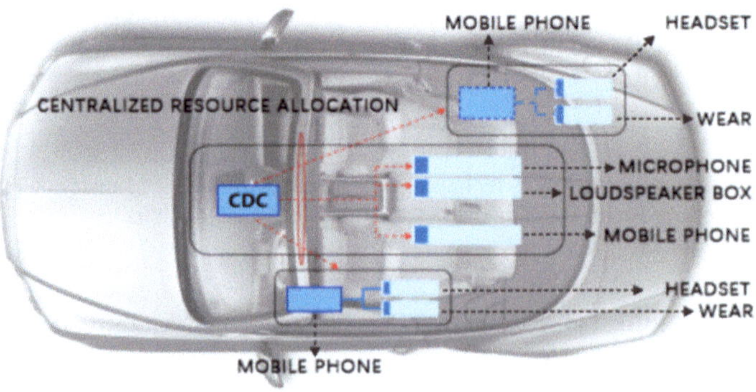

Fig. 1 Intelligent vehicle application

another communication domain. In the smart home scenario, the TV and the down-hanging audio device form one communication domain, and the mobile phone and earphone form another. The two communication domains can be distinguished by the advanced/common communication domain, and the advanced communication domain coordinates resources to achieve coordination and coexistence between multiple domains.

2.2 *Protocol Stack Architecture*

The system protocol stack is divided into application layer (OSI 5–7 layer), network and transport layer (OSI 3–4 layer) and access layer (OSI 1–2 layer), as shown in Fig. 2.

The data link layer ensures reliable data transmission. The data link layer consists of the link control layer and the media access layer. The link control layer mainly realizes transmission mode control, encryption and decryption, etc. The media access layer mainly realizes resource scheduling, data encapsulation, and transmission format control to meet QoS requirements of different services. The physical layer realizes the bit stream transmission function. The access layer also implements information security and management functions, which are used to ensure the security of the protocol stack and manage the communication. As shown in Fig. 3, in the process of data encapsulation, packet headers are added layer by layer at the data sending end. At the data receiving end, the data is unpacked in reverse order.

The system supports cross-layer transparent transmission for ultra-low latency periodic packet data transmission (such as audio transmission in active noise reduction services). This mechanism can determine the corresponding service parameters

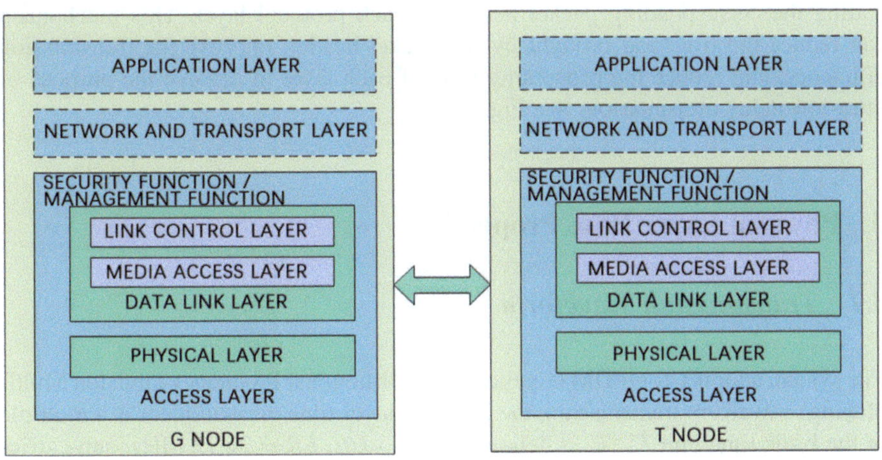

Fig. 2 System protocol stack architecture

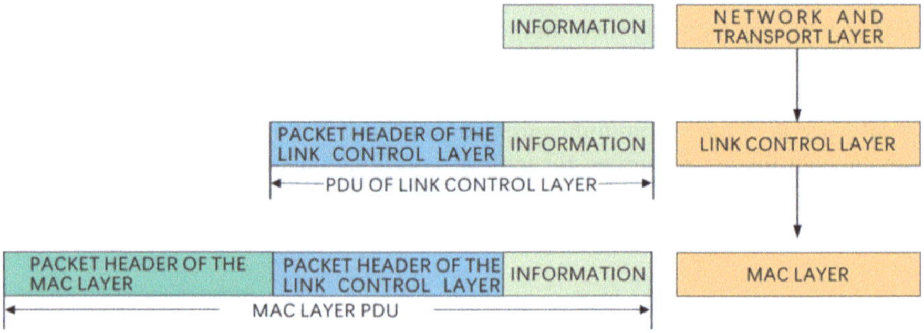

Fig. 3 System co-data encapsulation process

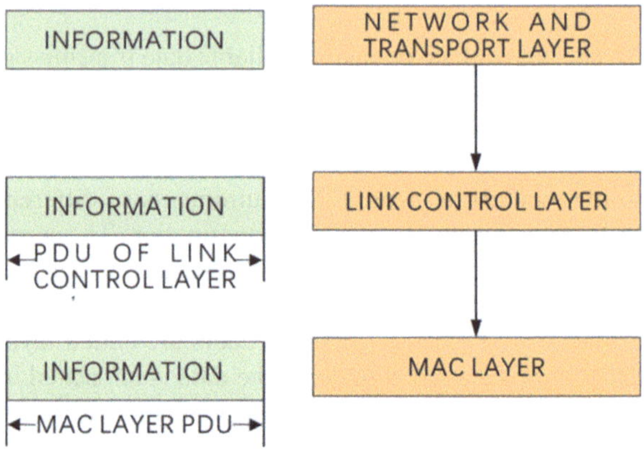

Fig. 4 System transparent transmission mechanism across layers

and the corresponding transport channel when the connection is established without adding the corresponding packet header at each protocol layer. This mechanism can reduce the overhead brought by the packet header, improve the transmission efficiency, and reduce the processing time of each layer to achieve the purpose of ultra-low delay transmission. See Fig. 4.

3 Physical Layer Base Properties

3.1 Transmission Waveform

The system adopts CP-OFDM (Cyclic Prefix-Orthogonal Frequency Division Multiplexing) waveform transmission, the physical layer time measurement is a multiple of the basic time unit T_s, T_s is defined as $T_s = 1/fs$, $FS = 30.72$ MHz, sub-carrier interval $\Delta f = 480$ kHz.

CP-OFDM symbol contains cyclic prefix part and valid data in the time domain. The length of the valid data part is $64T_s$, and the cyclic prefix length includes two types:

$$T_{\mathrm{CP}} = \begin{cases} 5 \times T_s, \text{Regular loop prefix} \\ 14 \times T_s, \text{Extended loop prefix} \end{cases} \tag{1}$$

CP-OFDM symbol time length (including the loop prefix):

$$\mathrm{T}_{\mathrm{Symb}} = \begin{cases} 69 \times T_s, \text{Regular loop prefix} \\ 78 \times T_s, \text{Extended loop prefix} \end{cases} \tag{2}$$

3.2 Channel and Subcarrier Design

The minimum carrier bandwidth of the system is 20 MHz, and the carrier bandwidth of 40/60/80/100/160/320 MHz is supported. The carrier bandwidth is composed of consecutive 20 MHz carriers. The carrier of 20 MHz is composed of 39 consecutive sub-carriers, with an interval of 480 kHz, numbered as #0,#1,... from the lowest frequency to the highest frequency.,#38, where the subcarrier #19 is the DC subcarrier and does not carry information. In a 20 MHz working bandwidth, the lowest frequency and the reserved part of the highest frequency resources do not place available subcarriers. See Fig. 5.

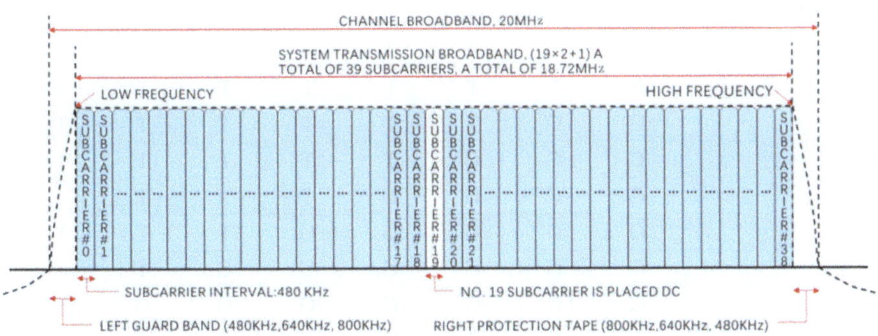

Fig. 5 System subcarrier division (20 MHz carrier)

3.3 Superframe Structure and Wireless Frame Structure

The system adopts TDD (Time Division Duplexing) mode, and the format of super frame is shown in Fig. 6 below. Each super frame contains 48 wireless frames, and the duration of each super frame is 1 ms, and the duration of each wireless frame is 20.833 μs. Where G symbol represents the symbol that G node sends to T node (G link), T node represents the symbol that T node sends to G node (T link), SG/ST respectively represents the symbol resource that can be used for overhead symbol in G/T symbol, and the overhead symbol resource of each wireless frame can be flexibly configured as 0, 1 or 2 symbols. GAP is the switching interval between G link symbol and T link symbol.

When using the conventional cyclic prefix, the wireless frame supports 14 G symbol T symbol ratio. When using extended loop prefix, wireless frame supports 12 G symbol T symbol ratio. Flexible G/T ratio can meet the requirements of service rates in different link directions in different application scenarios.As shown in Tables 1 and 2 below.

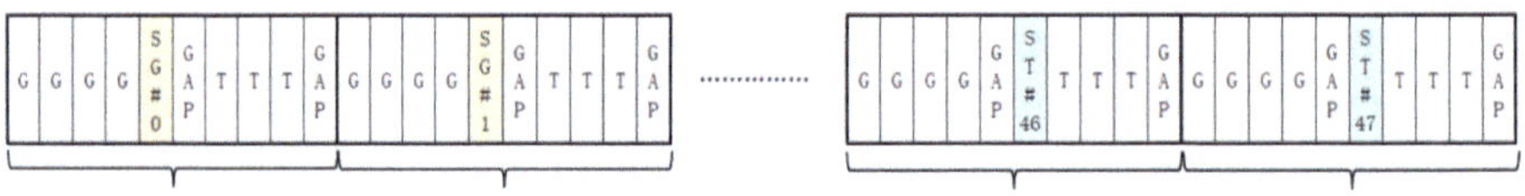

Fig. 6 System superframe structure

Table 1 Wireless frame ratio based on conventional cyclic prefix configuration

Wireless frame ratio	Symbol configuration							
	0	1	2	3	4	5	6	7
0	G	T	T	T	T	T	T	T
1	G	G	T	T	T	T	T	T
2	G	G	G	T	T	T	T	T
3	G	G	G	G	T	T	T	T
4	G	G	G	G	G	T	T	T
5	G	G	G	G	G	G	T	T
6	G	G	G	G	G	G	G	T
7	T	G	G	G	G	G	G	G
8	T	T	G	G	G	G	G	G
9	T	T	T	G	G	G	G	G
10	T	T	T	T	G	G	G	G
11	T	T	T	T	T	G	G	G
12	T	T	T	T	T	T	G	G
13	T	T	T	T	T	T	T	G

Table 2 Wireless frame ratio based on extended cyclic prefix configuration

Wireless frame ratio	Symbol configuration						
	0	1	2	3	4	5	6
0	G	T	T	T	T	T	T
1	G	G	T	T	T	T	T
2	G	G	G	T	T	T	T
3	G	G	G	G	T	T	T
4	G	G	G	G	G	T	T
5	G	G	G	G	G	G	T
6	T	G	G	G	G	G	G
7	T	T	G	G	G	G	G
8	T	T	T	G	G	G	G
9	T	T	T	T	G	G	G
10	T	T	T	T	T	G	G
11	T	T	T	T	T	T	G

4 Synchronization and Access

4.1 System Synchronization

The system is configured with two synchronization signals, FTS (first training signal) and STS (second training signal), which are placed in two adjacent wireless frames and sent in a cycle of superframes. ZC sequence is selected as the synchronization sequence. Compared with m sequencecommonly used as synchronization signal, ZC sequence has higher autocorrelation peak value, lower cross-correlation value and better anti-frequency offset performance, which is beneficial to improve the probability of synchronization success.

The FTS sequence function is:

$$d_{\mathrm{FTS}}(n) = \begin{cases} \exp\left(-j\frac{\pi u n(n+1)}{41}\right), & n = 0, 1, \ldots, 18 \\ 0, & n = 19 \\ \exp\left(-j\frac{\pi u n(n+1)}{41}\right), & n = 20, 21, \ldots, 38 \end{cases} \tag{3}$$

where, $u = 1$ for advanced communication domain and $u = 40$ for general communication domain.

The STS sequence function is:

$$d_{STS}(n) = \begin{cases} \exp\left(-j\frac{\pi u \frac{n}{2}\left(\frac{n}{2}+1\right)}{21}\right), & n = 0, 2, \ldots, 38 \\ 0, & n = 1, 3, \ldots 37 \end{cases} \tag{4}$$

Among them, $u = 1, 2, \ldots 20$ indicates the synchronization identifier of the communication domain. The STS sequence increases power by 3 dB when mapped to a subcarrier.

4.2 System Access and High Concurrent Transmission

High concurrency transmission mainly includes multi-node concurrency and multi-service concurrency. As shown in Fig. 7, multi-node concurrency means that a single G node can connect to multiple T nodes at the same time and provide services for multiple T nodes at the same time. Multi-service concurrency means that multiple types of services can coexist on a single T node to provide rich service experience for drivers.

To support high concurrency, the main technical points are as follows:

- Stable connection of a large number of users: In the system, the physical layer ID used to identify T nodes is 12 bits long. Theoretically, a SINGLE G node can support a maximum of 212 = 4096 T nodes.
- Access control mode: The system uses centralized scheduling to avoid link conflicts caused by preemption of distributed resources on a large number of nodes and improve system throughput. The system also supports non-competing access mode. A large number of T nodes can initiate group access from mutually orthogonal resources at the same time, achieving millisecond access and meeting service requirements in power-on and work-on scenarios.
- Intelligent scheduling based on service features: The system allows T nodes to report necessary service features to G nodes for intelligent scheduling. As shown in Fig. 8, for active denoising service, G node supports T node to report sampling

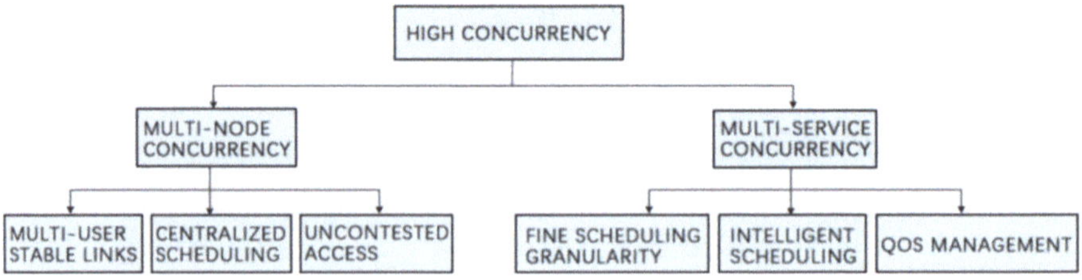

Fig. 7 High concurrency key technology

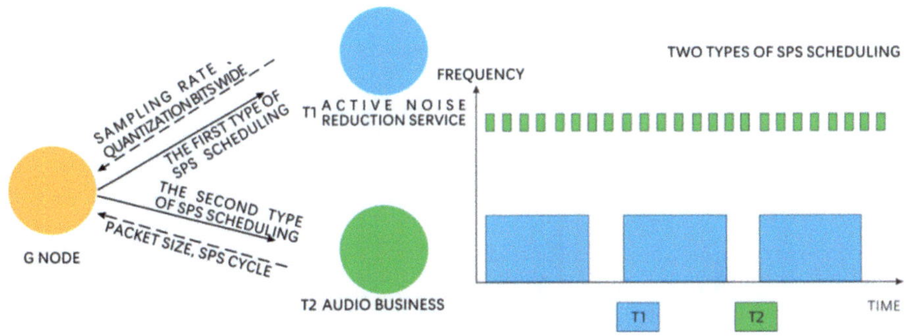

Fig. 8 System fine scheduling granularity and service intelligent scheduling

rate and quantization bit width. For semi-static scheduling services, T node can report the semi-static scheduling period and packet size to G node, facilitating G node to flexibly schedule services. In terms of scheduling, the system supports the logical channel priority mechanism to encapsulate data according to the logical channel priority. At the same time, the system can properly restrict the amount of data encapsulation of services with different priorities, and take into account the fairness of scheduling services with different priorities.

5 Resource Allocation

5.1 Resource Allocation in the Frequency Domain

The system supports ultra-low latency data transmission, also known as type 1 data transmission, and large packet/high traffic information, also known as type 2 data transmission.

Type 1 data transmission supports services with very low latency requirements. This type of data transmission requires high transmission reliability without retransmission. According to the transmission, system support minimum 1 is the carrier of scheduling granularity, G nodes according to the users in different channel fading of the subcarrier on (same subcarrier, different users to the corresponding frequency domain decline coefficient), scheduling of different sub-carrier data transmission, realize the combination of sub carrier scheduling, to the greatest extent promote each user and the system performance. As shown in Fig. 9 below, a single user is taken as an example to schedule the subcarrier with optimal channel conditions to transmit data.

For the second type of data transmission, the system supports scheduling granularity with 4 or 3 subcarriers as a group, so as to obtain sufficient frequency domain scheduling gain while reducing signaling indication overhead.

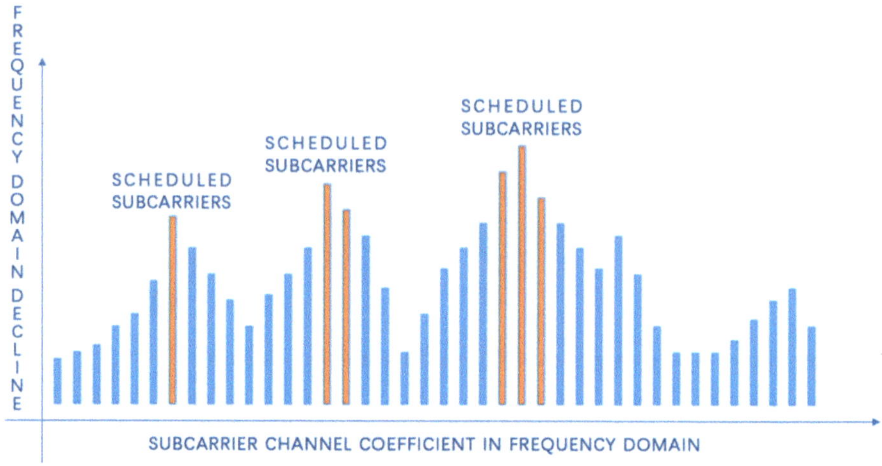

Fig. 9 Schematic diagram of decentralized scheduling principle

For system overhead signals, such as T-link ACK feedback signals and T-node access information, the system supports frequency resource comb transmission and the frequency resources used in information transmission are distributed within the entire bandwidth of a carrier to ensure the maximum frequency diversity of information transmission.

5.2 *Time Domain Resource Allocation*

With 20.833 μs wireless frame as the scheduling unit, the system supports flexible resource allocation in the time domain to meet the delay requirements of different application scenarios. The following key technologies are used to allocate time domain resources in the system:

- Short wireless frame: The length of each wireless frame is 20.833 μs. Because G link and T link can be configured for wireless frame transmission at the same time, the one-way transmission delay at the physical layer is no longer than 20.833 μs. The following Fig. 10 schematically analyzes two scenarios in the case of semi-static scheduling with G link as an example: Scenario 1 (Case 1), the fourth symbol of each wireless frame is used for data transmission. If the data packet is ready at the first symbol, the data packet can be sent only after 3 symbols, with the transmission delay less than 20.833 μs, because the latest resource available for transmission is at the fourth symbol. In Scenario 2 (Case 2), assuming that the packet is ready after the third symbol of the current wireless frame, the packet is sent at the fourth symbol of the next wireless frame, and the longest transmission delay of the packet is only 20.833 μs.

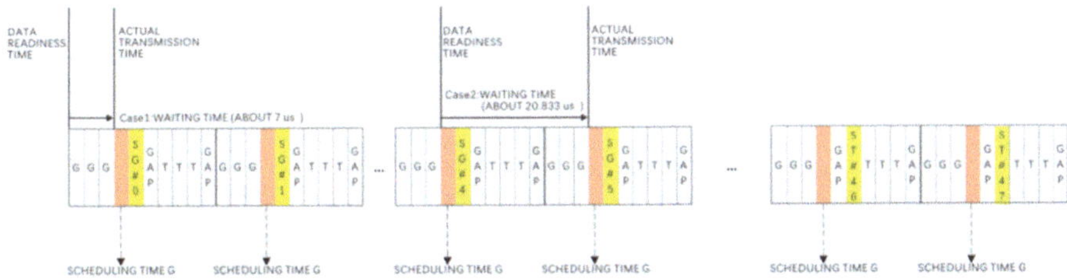

Fig. 10 Packet sending diagram

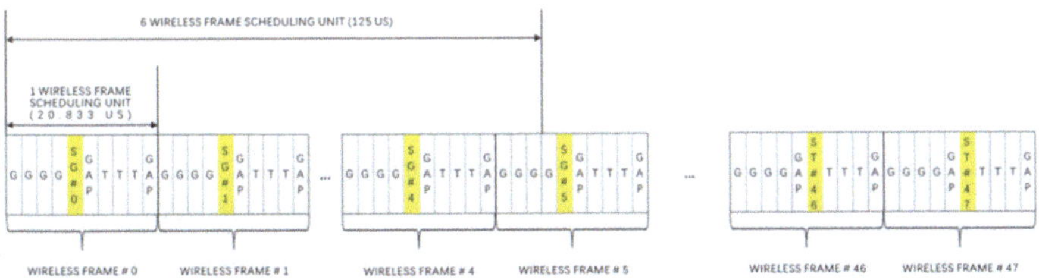

Fig. 11 Schematic diagram of packet scheduling unit (period)

- Overhead resources evenly distributed signal, as shown in Fig. 10, overhead signals (such as sync signal, broadcast information, control information, access to information, ACK feedback information, etc.) the physical resources across multiple wireless transmission in the frame, to ensure that each wireless frame has resources transmission G link packet and a link.
- Ultra-short flexible scheduling period: The star flicker system supports two types of scheduling unit (period) configurations. Ultra-low latency data transmission supports the shortest scheduling period of wireless frame duration (20.833 μs). Each wireless frame contains at least one transmission opportunity of G link and T link, and the minimum transmission delay is one wireless frame duration (20.833 μs). Large packet/high traffic data transmission supports a minimum unit of 6 wireless frame scheduling cycles (125 μs). See Fig. 11.

6 Channel Coding and Modulation

6.1 Data Transmission Channel Coding

Polar code is a channel code constructed on the basis of channel polarization theory. It is a channel code that can reach Shannon limit after theoretical analysis and demonstration, and can better resist random errors. RS code is a linear block code, which is based on the multi-base channel code of Galois Field. Each symbol can contain

Table 3 Ultra low delay data transmission modulation coding

Bits	Encoding block length (including 8 bits CRC)	Channel coding scheme	Modulation method
16bits	24bits	RS (15,11) truncated (10,6) or Polar	QPSK
			16QAM
			64QAM
			256QAM
			1024QAM
24bits	32bits	RS (15,11) truncated (12,8) or Polar	QPSK
			16QAM
			64QAM
			256QAM
			1024QAM
32bits	40bits	RS (15,11) truncated (14,10) or Polar	QPSK
			16QAM
			64QAM
			256QAM
			1024QAM

more than one bit. It has good anti-burst interference performance and can better resist continuous errors. The system uses Polar code or RS code to transmit small packet services with ultra-low delay (such as vehicle-mounted active noise reduction) to ensure that the system can achieve highly reliable transmission in different application scenarios. The specific coding and modulation combinations are shown in Table 3.

6.2 *The Physical Layer HARQ*

Hybrid ARQ (HARQ) is a combination of FEC and ARQ to increase the transmission reliability of links. In traditional ARQ, when the receiving end detects an error in the received information, the receiving end directly discards the error packet and requests the sending end to retransmit the corresponding packet. Compared with ARQ, HARQ enhances ARQ. That is, the received error packet information is not discarded, but is combined with the retransmitted packet information to improve the receiving reliability.

The system adopts Polar code based asynchronous HARQ technology, supports up to four HARQ processes, and supports CC-HARQ scheme and IR-HARQ scheme. The benefits of CC-HARQ scheme come from multiple soft information merging at the receiver, which improves the equivalent SNR of the information at the receiver and reduces the error probability. In IR -HARQ scheme, according to the characteristics

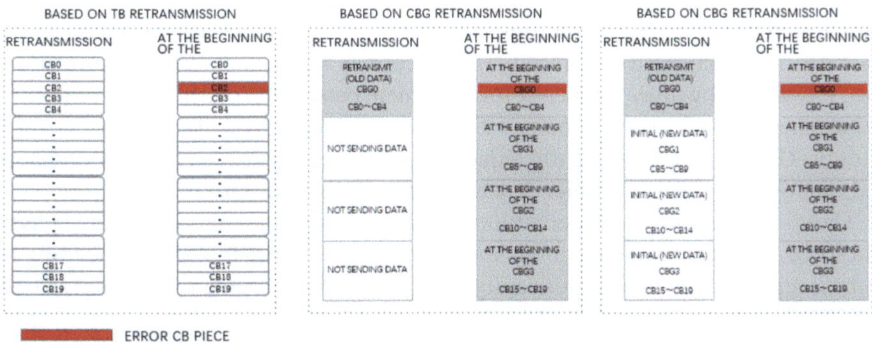

Fig. 12 HARO scheme

of Polar code, the length of the master code is extended during retransmission or the encoding bits not sent during the first transmission are sent, and the encoding gain is further obtained on the basis of the energy gain.

The system supports three retransmission schemes: retransmission based on TRANSFER block (TB), retransmission based on code block group (CBG) and mixed retransmission based on CBG.

TB retransmission means that if any CB in a TB fails, the data in the entire TB is retransmitted. During the retransmission, the number of CB segments C is the same as that in the first transmission, but the channel bits of each CB may be different from those in the first transmission.

CBG retransmission for each CB, the implementation process is the same as TB retransmission. Different from TB retransmission, CBG retransmits only the CBG where the CB error occurs.

CBG mixed retransmission refers to the CBG retransmitted by the previous TB and the CBG initially transmitted by the new TB. Segment number C is the same number of block segments contained in the last TB initial transmission associated with this transmission. All incoming CBGS form a new transport block (TB) (Fig. 12).

7 Multi-Domain Collaborative

7.1 Multi-Domain Synchronization

The system reduces interdomain interference through time/frequency synchronization between multiple G nodes. The star flicker system adopts OFDM waveform. In the scenario where multiple communication domains exist, even if different communication domains use different frequency points, if the frequency difference is not an integer multiple of the sub-carrier interval SCS = 480 kHz or the timing difference exceeds CP, interference between sub-carriers will be caused. In particular, when the interference comes from multiple communication domains and the interference source is much closer to the receiving device than the signal source, the interference introduced by time–frequency misalignment between G nodes can significantly reduce the received signal-to-noise ratio. Time/frequency synchronization between multiple G nodes can significantly reduce the interference between multiple domains and improve the spectral efficiency when multiple domains coexist.

The system needs to consider multiple communication domains in the same physical space. In a dense deployment scenario, the path loss from the interference source to the receiver may even be significantly smaller than that from the signal source to the receiver. Taking two communication domains using adjacent carriers as an example, considering the same transmitting power of G nodes in the two domains, the path loss from the interference source to the receiver from other domains is 20 dB smaller than the path loss from the signal source to the receiver. When the two domains are synchronized, and the frequency synchronization error is 100 Hz, for example, the power leaked from the interference source to the carrier relative to the interference source is less than −74 dB, the received signal dry ratio is higher than 54 dB (74–20), and the interference can be ignored. When the two domains are asynchronous, generally, only filters can be used to suppress the adjacent frequency interference. The power leaked from the interference source to the carrier is about 20 dB relative to the power of the interference source, and the received signal dry ratio is about 0 dB (20–20). It is not difficult to see that time/frequency synchronization between multiple G nodes can significantly reduce the interference between communication domains, especially the interference between communications domains located on adjacent carriers.

Time/frequency synchronization between multiple domains and multiple G-nodes can be completed by sending synchronization information and listening behavior of g-nodes. Figure 13 takes five G nodes as an example. The startup sequence is G1, G5, G4, G2, and G3, showing the process of realizing time–frequency synchronization in five domains.

- First of all, G1 and G5 nodes are not overwritten by any synchronization set when they start up, so the red synchronization set and blue synchronization set are established respectively.
- Then, with the startup of G2 node and G4 node, the number of nodes in red synchronization set and blue synchronization set gradually increases, and the coverage is gradually expanded to cover G3 node.
- Next, when G3 is powered on, the red synchronization set and blue synchronization set are listened to, and both synchronization sets have two nodes. G3 is closer to G4 node, so the blue synchronization set has stronger coverage on G3, and G3 is added to the blue synchronization set.
- At last, G2 nodes and G1 nodes are successively transferred from the red synchronization set to the blue synchronization set with more nodes. Finally, all nodes are in the blue synchronization set and the synchronization is completed.

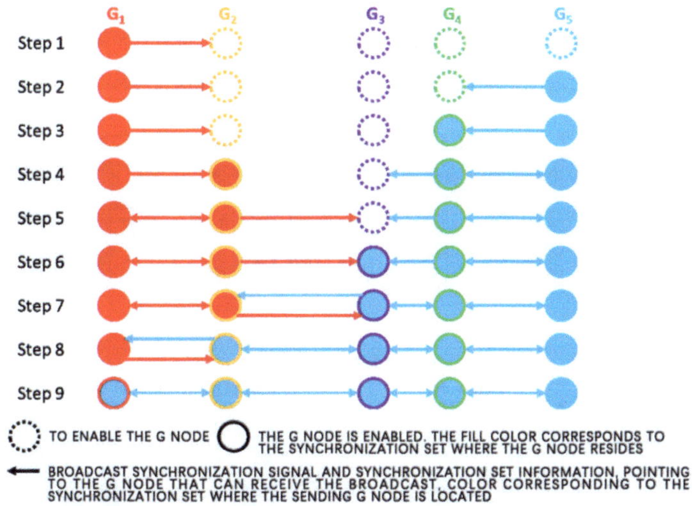

Fig. 13 Schematic diagram of multi-node time–frequency synchronization process

7.2 Multi-Domain Resource Coordination

The system supports G nodes in advanced communication domains to allocate resource pools to other communication domains through broadcast to achieve multi-domain resource coordination. Resource pools in different communication domains can use different carriers, and resource pools in different communication domains on the same carrier can use different symbols. On each carrier containing the advanced communication domain, the G node of the advanced communication domain instructs different communication domains to use the resource pool with orthogonal time domain through the system message, so as to avoid the communication links of different communication domains using the same resource.

Figure 14 shows an example of resource usage over two communication domains on the same carrier. In the resource pool of the communication domain:

- The resources used for data transmission (G symbol and T symbol in the figure) are allocated in symbol granularity and repeated in wireless frame cycle.
- The resources used for overhead transport (the S symbol in the figure) are allocated at overhead symbol granularity, repeating in superframe cycles.

When the resource pools of different communication domains are different, G node of the communication domain sends synchronization signals and transmits service data in the resource pool of its own communication domain. Then, the G node can receive synchronization signals of other domains in the resource pool of other communication domains to achieve inter-domain synchronization tracking, without affecting service transmission in the communication domain.

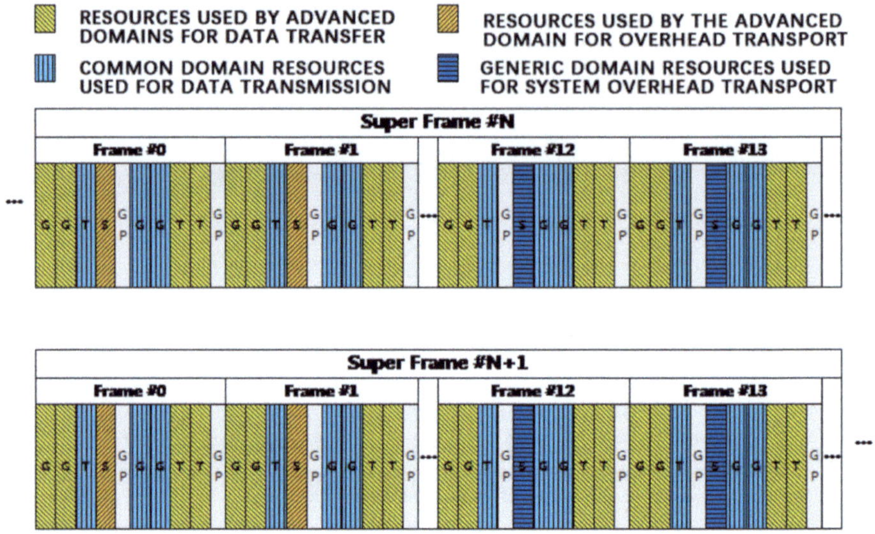

Fig. 14 Diagram of multi-domain resource coordination

8 Conclusion and Outlook

In conclusion, the system can satisfy the ultra-low delay application in vehicle-mounted application scenarios by defining ultra-short timeslot frame structure and ultra-short wireless frame scheduling period (20.833 μs). High-performance channel coding, physical layer HARQ retransmission and discrete single carrier scheduling technologies can achieve highly reliable transmission, meet the requirements of 99.999% or more highly reliable applications, and realize the replacement of terminal wired connections. Multi-domain collaboration technology improves resource efficiency and reduces interference between networks. Minimalism and high security information features are designed to meet the needs of high security applications.

Future system will be at a higher efficiency, lower power consumption, larger bandwidth, more antennas and other technical direction for continuous evolution, and support range, more functions, such as networking launch with higher speed, lower cost, lower power consumption, and further the coverage of the evolution of the system, better support all kinds of application scenarios, in short distance wireless applications play a bigger role.

Acknowledgements This work was supported by the Central Public Research Institutes Special Basic Research Foundation (2021-9013a).

References

1. Zhu, S.L.: Thoughts on the development of car ownership. Shanghai Auto **2018**(2), 24–26 (2001)
2. Kato, K., Suzuki, M., Fujita, Y., et al.: Image synthesis display method and apparatus for vehicle camera: US 7139412, 11–21 (2006)
3. Lu, B., Qin, R., Li, Q., Chen, D.P.: Study of vehicle-surrounding image stitch algorithm. Comp. Sci. **40**(9), 293–295 (2013)
4. Luo, L. B., Koh, I. S., Min, K. Y., et al.: Low-cost implementation of bird's eye view system for camera-on-vehicle. In 2010 Digest of Technical Papers International Conference on Consumer Electronics (ICCE). IEEE, 311–312 (2010)
5. Cancare, F., Bhandari, S., Bartolini, D. B., et al.: A bird's eye view of FPGA-based Evolvable Hardware. In 2011 NASA/ESA Conference on Adaptive Hardware and Systems (AHS). IEEE (2011)

Countermeasure Research and Application of Beijing-Taipei Expressway Based on Variable Message Signs

Yuchen Liu, Tao Li, and Hengbo Zhang

Abstract Variable Message Signs (VMS) is one of the important components of Advanced Traveler Information Systems, which provides drivers with dynamic traffic guidance Information of the road ahead. Drivers are guided to choose the best travel route, to avoid the congested sections and optimize the operation of the road network, to realize the reasonable distribution of traffic flow in the road network. Based on the project of The Beijing-Taipei Expressway (Shandong section), this paper summarizes the existing service level of the Beijing-Taipei Expressway, summarizes, and analyzes the existing problems of the variable message signs, and on this basis, puts forward countermeasures and suggestions for the variable message signs to improve travel efficiency and enhance user satisfaction.

1 Introduction

The Shandong section of Beijing-Taipei Expressway is in the middle and west of Shandong province, running through the north and south of Shandong province, connecting dezhou, Jinan, Tai 'an, Jining, Zaozhuang and other 5 large and medium-sized cities and 17 counties and cities along the route. It is an important part of Shandong province's expressway network layout plan "nine vertical, five horizontal, one ring, seven links". The reconstruction and expansion project are to improve the national and provincial highway network, to meet the needs of the development of the national comprehensive transport channel, but also to improve the highway capacity and service level, is to implement the national and provincial regional strategy, It plays a positive role in speeding up the construction of modern comprehensive transportation system, strengthening the construction of "safe transportation" and promoting the development of regional tourism resources and tourism.

Y. Liu (✉) · H. Zhang
Beijing GOTEC ITS Technology Co., Ltd, Beijing 100088, China
e-mail: lyc@itsc.cn

T. Li
Shandong High-Speed Construction Management Group Co., Ltd, Jinan 250101, China

L. C. Jain et al. (eds.), *Advances in Wireless Communications and Applications*,
Smart Innovation, Systems and Technologies 299,
https://doi.org/10.1007/978-981-19-2255-8_16

At present, the traffic service mode of Beijing-Taipei Expressway can be roughly divided into Application terminal, variable message signs, vehicle-mounted terminal, traffic broadcast and so on. Among them, the most frequently used information service means is APP terminal, which can use mobile phones, portable android device, and other ways to obtain traffic information through the communication network. It is simple and convenient to use, without extra cost, and can obtain traffic information anytime and anywhere. However, the usage rate of most APP terminals is only about 70%.

The variable message signs are suitable for releasing variable information, as well as weather information and advertising. Drivers and passengers can intuitively understand the conditions of the road ahead, and its service objects are mainly motorists. The variable message signs have the advantages of intuitive performance and strong system stability. However, the existing information boards have the disadvantages of poor pertinence, users can not actively obtain information, the obtained information cannot be stored, and the construction cost is high. To better improve the traffic information service capacity of Jingtai Expressway and explore the way to obtain higher user satisfaction, this paper studies countermeasures based on variable message signs [1].

2 Beijing-Taipei Expressway Service Level

The Shandong section of the Beijing-Taipei expressway, located on a major highway in Shandong province, is expected to see more than 40,000 traffic per day by 2023 and more than 86,000 by 2047. According to the analysis of work availability report, the proportion of vehicle types in each section of the project is predicted as follows (Table 1):

At present, passenger transport will continue to increase, the proportion of small buses will rise. After the opening of relevant railways in the region, part of long-distance passenger traffic will be diverted, and the proportion of buses will decrease year by year. Because the transport efficiency of large trucks is significantly higher

Table 1 Forecast table of the proportion of future vehicles on the project Road

Year	Van (%)	In the van (%)	Big trucks (%)	Heavy trucks (%)	Container (%)	Car (%)	Bus (%)	All (%)
2023	7.10	3.80	5.60	21.60	2.80	56.40	2.70	100.00
2025	7.00	3.80	5.70	21.60	2.90	56.40	2.60	100.00
2035	6.60	3.70	5.90	21.70	3.10	56.60	2.40	100.00
2040	6.40	3.70	6.00	21.80	3.20	56.70	2.20	100.00
2042	6.30	3.70	6.00	21.80	3.30	56.70	2.20	100.00
2045	6.20	3.60	6.10	21.90	3.40	56.80	2.00	100.00
2047	6.10	3.60	6.10	21.90	3.50	56.80	2.00	100.00

than that of medium and small trucks. Accordingly, the proportion of large trucks such as large trucks, extra-large cargo and containers will increase while that of small and medium cargo will decrease. Due to the large volume of traffic, the service means currently designed is difficult to meet the needs of users, so it is necessary to study better service means on this basis.

3 Analysis of Traffic Guidance Status of Variable Message Signs

According to the survey of Highway Travel Service conducted in the early stage, the analysis shows that at present, the way for travel users to obtain road information is navigation software, accounting for 74.29%, and the attention to road signs and LED display, accounting for 16.64%. Users' attention to WeChat public accounts (Shandong Expressway, Shandong Expressway Traffic Police, etc.) is 5.55%, followed by FM traffic broadcasting and website users (microblog, official website) (Fig. 1).

It can be seen from the above chart that the variable message signs accounts for less than 20% of users' access to road information. Through analysis, the reasons are as follows:

3.1 The Content of Variable Intelligence Board is Inaccurate and Lacks Real-Time

The variable message signs provide real-time traffic information to the driver, and the driver receives and responds to the induced information, to achieve the purpose of rational distribution of traffic flow [2]. In the process of the driver's path choice decision, the road condition information displayed by the variable message signs

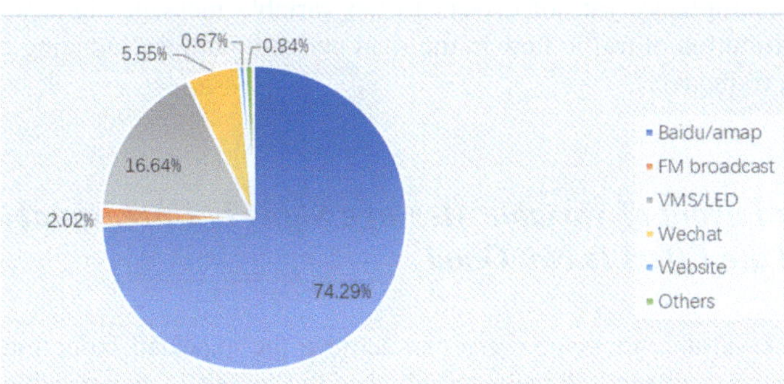

Fig. 1 Proportion of the way that travel users get road information

has a great influence on the driver's path choice behavior. If the road condition information displayed by the variable message signs is accurate, the driver will make an appropriate path choice according to the information. On the contrary, drivers' choices may lead to more congestion on congested roads and worse road network performance. But the driver in the process of driving, his route choice behavior is not only related to the accuracy of the content of the variable message signs, more associated with the real-time performance of the variable message signs information, from the collection network status to pass information to the driver, from the pilot receives information to respond according to the information needs to be a certain reaction time, and the path of the congestion status may change at any time.

3.2 *Variable Message Signs Is Unreasonable and Counterproductive*

The compliance rate of drivers to the variable message signs has a direct impact on the distribution of traffic flow in the road network and has a very important impact on the operation and management of the road network. But the driver on the size of the variable message signs induction information compliance rate has always been difficult to determine, on the one hand, the driver itself to variable message signs trust in a certain extent, affected the compliance rate of the variable message signs induction information, the driver to the variable message signs trust is higher, the more likely they were to change their travel path according to the variable message signs. Impact driver for variable message signs, on the other hand, inducing factors of information compliance rate is varied, the driver's perception of the variable message signs kimono properties affect the driver's behavior, the driver's travel properties may also affect its decisions, even the driver's personal attributes to a certain extent, also affect the route choice behavior [3]. Moreover, the public's recognition of the variable message signs continues to improve, and the size of the driver's compliance rate to the variable message signs induced information will also change. Too high or too low compliance rate of drivers to the variable message signs will lead to uneven distribution of traffic flow in the road network, and the inducing effect will be counterproductive.

3.3 *The Layout of Variable Message Signs Is Unreasonable, and the Effect Is Not Good*

Whether the variable message signs can achieve the expected induction effect in the road network depends largely on whether the layout of the variable message signs in the road network is reasonable. Induced effects, to maximize the variable message signs variable message signs layout position to keep proper distance with the driver, otherwise, the pilot induced information has been received too late diverted into congestion area, and the distance is too large, the driver in advance to choose diversions of traffic, the unobstructed replacement path have new congestion, and originally congested roads has returned to normal, The variable message board cannot achieve the desired induction effect. Furthermore, at present, most variable message signs are mainly arranged on urban expressways, and the induced information provided is mostly the road condition information of expressways, while the variable message signs are basically not arranged on other paths. The road network information obtained by drivers is one-sided and incomplete, leading to drivers unable to make correct decisions according to the variable message signs induced information.

4 Research on Standards and Specifications

Follow the recommendations in the traffic information issued by the Road Traffic Information Service through the variable intelligence board in GB-T 29103-2012. Traffic information can be classified in different ways as follows:

- According to the timeliness of traffic information is divided into dynamic information and static information. The details are as follows (Table 2)

Table 2 Information classification

Information classification	Describe objects	The specific content
Dynamic information	Traffic information	Ice, water, snow, congestion, traffic volume, average speed, etc.
	Meteorological information	Wind, visibility, estimated duration, etc.
	Control information	Tunnel closure, lane closure, estimated duration, etc.
	Event information	Location, size, cause, severity, expected duration, etc.
	Response	Detour route, speed limit, etc.
	Traffic information	Ice, water, snow, congestion, traffic volume, average speed, etc.
Static information	Basic Tunnel Information	Length of tunnel, number of tunnel lanes, distance to entrance/exit, etc.
	Traffic restriction information	Limit height, width, speed limit, articles, etc.
	The service information	Rescue, consultation, emergency calls, etc.

- According to the form of release, it can be divided into text information, graphic information, and image information.

In the figure, green is used for smooth driving areas, yellow is used for slow driving areas, and red is used for congested areas. The image information resolution should be no less than 320 × 200, the image refresh rate should be no less than 10 frames per second, and the image continuous playback time should be no less than 5 s. Image traffic information should not be used on the variable message signs.

Through the compilation of instructions to view, through the video image dynamic release of some complex information, has the characteristics of comprehensive and diverse information, comprehensive expression, large amount of information, rely on the existing technical means to achieve image information release, need to have relatively high-cost investment, and the release of images have certain restrictions. So, these uses are regulated.

5 Study on Traffic Guidance Countermeasures of Variable Message Signs

5.1 Real-Time Variable Message Signs Display Content

The traffic guidance information displayed by the variable message signs plays a role by influencing the driver's path choice behavior, and its effect depends on the driver's response to the information [4]. Since the driver's behavior is active and conscious rather than passive, the driver's opinion of information accuracy will affect their trust in information. Therefore, the accuracy of traffic information determines the reaction behavior of drivers. High-quality traffic information can achieve good results, while low-quality traffic information not only cannot induce traffic well, but even lead to the weakening of drivers' trust in information, affecting drivers to take correct and reasonable driving behavior. To ensure that the variable message signs can provide accurate and real-time induction information, it is suggested to combine the data of ETC portal frame with the cooperation of Internet companies to do real-time road condition analysis of traffic big data [5]. In addition, based on the existing historical data of Internet companies, the traffic congestion in special periods such as peak hours and holidays can be analyzed in advance, and the model data algorithm of cloud can be used to make traffic guidance in advance. Besides, the key road information should be highlighted, such as font width or flashing, to attract the driver's attention. In addition, the display of day and night should be different to accommodate the difference in light.

5.2 *Front-End Awareness Device and Event Release*

Taipei high-speed reconstruction after, there will be a lot of intelligence equipment installation, these sensors to upload data and alarm information, the wisdom of the highway platform and third-party system event information can be released on the variable message signs linkage, is different from the traditional artificial release induced traffic information, improve the variable message signs information release ability. The types of warning messages that can be issued jointly include congestion event, rescue event, accident event and road-related construction event; Temperature and humidity alarm and visibility alarm uploaded by temperature and humidity sensor and visibility detector; Video detection events reported by cameras, such as pedestrians, vehicles going backwards, objects thrown on the road, construction, and other video detection events; Congestion alarm provided by radar congestion detection. As information sources, the above event types can be used to realize fully automatic and full-process intelligent traffic guidance in a real sense with the help of fully automatic information board publishing technology.

5.3 *Reasonably Variable Message Signs Compliance Ratio*

The obedience rate of drivers to the inducible information of the variable message signs largely depends on the driver's correct perception of the congestion degree of the path displayed by the variable message signs. The color classification is too bright for the traffic congestion situation displayed by red, yellow, and green colors. It is suggested to add multiple colors. In addition, due to the different driving experience and culture, the perceived road congestion degree is also different. In addition, drivers also want more detailed information about the cause, duration and, if possible, real-time video of traffic conditions so that they can make more accurate decisions based on that information. Finally, it is also important to construct a route selection model that considers drivers' personal attributes, travel attributes, cognition and obedience to the variable message signs, and even environmental attributes for studying drivers' obedience rate to the variable message signs.

5.4 *Standardize the Display Form of Variable Message Signs*

Text variable message signs can clearly convey information to the driver, but the display space is limited. The variable form of the graph is more informative and easier to understand. In difficult visual conditions (such as low brightness, fast display, etc.), graphic symbol information is superior to text information both in recognition speed and distance. Another advantage is that graphics are not limited by language, if the design pattern image, intuitive, different countries, different nationalities, different language drivers can understand and read. However, graph information requires drivers' higher attention and longer reading distance. The layout of the variable message signs with text and text is more complex, and the traffic state is displayed by graphics, while the text is simply supplemented. Whether it is text or graphics, it is generally matched with the corresponding color to warn the driver. Green indicates that the road is smooth, yellow indicates that the road is moving slowly, and red indicates that the traffic is congested. When the driver cannot see all the information of Chinese characters clearly, he can make a corresponding decision on the road condition according to the graphic information and different colors. Text variable message signs is suitable for displaying clear instructions or releasing simple information. The upstream of the main entrance ramp is released in the form of text, and the main line is released in the form of graphics, and the text and text mixed variable message signs is suitable for laying in the upstream of the main road exit.

5.5 Planning the Location of the Variable Intelligence Board

Only by reasonably planning the placement position of the variable message signs and reserving the appropriate reaction time for the driver can the variable message signs achieve the expected induction effect. First, determine the traffic node where the variable message signs needs to be set, and determine the final position of the variable message signs according to the road environmental conditions of the section where the node is located, combined with other static traffic signs, and then according to the external conditions of the road condition [6]. Furthermore, the appropriate distance between the position of the variable message signs and the driver should be reasonably determined according to the different situations of the induced information category of the variable message signs. Drivers should be informed in advance of accidents, construction and other congestion information that lasts for a long time, leaving enough time for drivers to change lanes to avoid congestion; However, for general traffic congestion information, it is not necessary to inform drivers too early, to avoid many drivers choosing to change routes and forming new congestion on alternative routes. In addition, the number and density of variable message signs should be increased, not only on urban expressways, but also on urban trunk roads, or even on branch roads, to provide more information about congestion paths and alternative paths, to facilitate drivers to make correct decisions.

6 Conclusions

This paper studies the countermeasures of variable message signs under the condition of standard specification. Variable message signs is not only a kind of traffic information release mode, but also a kind of traffic guidance control system [7, 8]. Reasonable variable message signs obedience ratio, guide the driver to choose the best path; Standardize the display form of variable message signs to meet the regional needs of different drivers; The location of the variable message signs is planned, and reasonable reaction time is reserved for drivers, to provide drivers with more traffic guidance information. The research results guide drivers to obtain effective traffic information, reduce the probability of frequent congestion and congestion on the Beijing-Taipei Expressway, enhance the level of traffic service informatization, improve the quality of public travel service, and enhance the image of public service. Due to the limited funds and time of the project, this paper only analyzes the countermeasures of the variable message signs. Specific data analysis, algorithms and strategies are also key steps to improve the ability of variable information service. To achieve efficient and timely traffic guidance and improve traffic, further research in these aspects should be carried out in the later stage.

References

1. Tian, F., Liu, Y.C., Cai, L.: Research and application of tracking service based on variable message signs. Highway Transportation Technology (Application Technology Edition) **16**(183(03)), 319–322 (2020)
2. Guo, Y.R.: Research on short-term Traffic Flow prediction and guidance method based on urban Road Traffic Data. Lanzhou University of Technology (2021)
3. Cao, Y.K.: Analysis and modeling of driver's path choice behavior under VMS. Traffic Inform. Safety **6**, 96–101 (2016)
4. Wei, K.: Expressway intelligent traffic guidance system based on GIS and information board. Western Transportation Science Tech. **167**(6), 165–168 (2021)
5. Peng,Z.B., Wu, Y.P., Rong, J., Xu, C.: Research on threshold of road node number based on variable portal information board. Traffic Information and Safety **38**(2), 55–132 (2020)
6. Tang, Z.: Status and Development of variable Traffic Information Signage (VMS) technology. China Municipal Engineering **03**, 13–15 (2015)
7. Cheng, S., Li, Q., Xu, C., Yao, J., Li, X.F.: Generalized vehicle-road cooperative traffic control system on expressways. Beijing: CN112907952A, 2021-06-04 (2021)
8. Xu, H., Yang, S.X., Yang, Z.: Research on information release content of Shanxi Expressway variable information board under event state. China Traffic Inform. **215**(3): 88–91 (2018)

Research on Spatio-Temporal Publishing Strategy of Traffic Guidance Information

Tao Li, Hengbo Zhang, and Qun Liu

Abstract Scientific and reasonable induced traffic information release strategy can effectively reduce traffic congestion, this paper first research and put forward the general structure of information service system, and then further study generalized information service oriented induced traffic information release strategy of time and space, based on traffic wave theory, put forward the total traffic incident duration and accident influence scope and traffic event queue length prediction model, Through the induced traffic information release area, put forward many terminal induced traffic information release strategy, and finally analyzed the relevant verification, analysis results show that the information release strategy can effectively guide the induced traffic information release is scientific and certainty, further enhance the highway public travel information service level.

1 Introduction

With the high-speed development of the highway informatization, the domestic highway development trend of information release way also is diverse, such as car terminal, VMS (Variable Message Sign), mobile phone APP (Application) and website, etc. Although diversified ways to disseminate information to a certain extent, improve the highway public travel information service level, but the information release system are independent. Moreover, China's expressway guidance information release system needs to be improved. Traffic incident information decision and release are mostly rely on managers' experience for manual operation. It is difficult to provide active, personalized and refined traffic information services to travel users because a complete publishing strategy of road guidance information has not been formed yet [1].

T. Li · Q. Liu
Shandong High-Speed Construction Management Group Co., Ltd., Jinan 250101, China

H. Zhang (✉)
Beijing GOTEC ITS Technology Co., Ltd, Beijing 100088, China
e-mail: zhb@itsc.cn

L. C. Jain et al. (eds.), *Advances in Wireless Communications and Applications*,
Smart Innovation, Systems and Technologies 299,
https://doi.org/10.1007/978-981-19-2255-8_17

2 Structure of Generalized Information Service System

Based on the existing travel information service publishing modes, the highway guidance information publishing strategy oriented to generalized information service is studied. Establish effective traffic guidance information release strategies for different information release systems such as VMS, vehicle-mounted terminals, and mobile phone apps through scientific methods, and provide travelers with real-time, active, and personalized traffic information services [2]. The system architecture is shown in the Fig. 1 below:

The system includes information acquisition layer, information processing layer and information release layer. The information acquisition layer mainly obtains traffic related information such as traffic flow, congestion state, event type and weather condition through information acquisition equipment. Followed by information processing layer, because of the collected data format and standard not unified problems, needs the data cleaning, after processing the data in the database of the standby, in addition, the collected data needs to be further data filtering, data fusion, data mining and data retrieval process. Then through the related algorithm and calculation model, the prediction model of the spatio-temporal range of traffic events and the information release area were obtained. Combined with the traffic related information such as the event type and road network operation state, the spatio-temporal

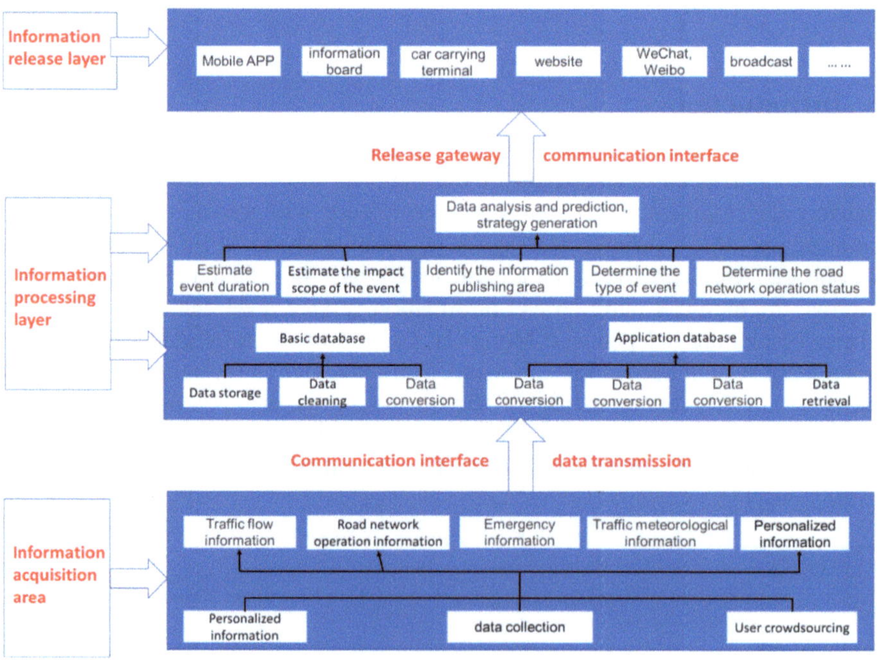

Fig. 1 Generalized information service system architecture

release strategy of induced information was formed [3]. The information release layer will release the processed data through mobile phone APP, VMS, vehicle-mounted terminals, websites, WeChat public accounts, broadcasting, and other publishing methods [4].

3 Spatio-Temporal Publishing Strategy of Traffic Guidance Information for Generalized Information Service

3.1 Prediction Model of Total Duration of Traffic Events

T1 prediction model of traffic accident duration. This study defines the duration of the traffic accidents as dependent variable, the factors that will affect the traffic incident duration is defined as the independent variable, the duration of traffic accident is defined as the dependent variable, assuming a linear relationship between the dependent variable and each variable, get type (1):

$$T_1 = \beta_0 + \beta_1 X + \beta_2 X_2 + \beta_P X_P + \varepsilon \tag{1}$$

where, $\varepsilon \sim N(0, \sigma^2)$, $\beta_0, \beta_1, \beta_2 \ldots, \beta_P, \sigma^2$ is unknown, and matrix is used to obtain the formula (2):

$$T_1 = \begin{Bmatrix} y_1 \\ y_2 \\ \cdots \\ y_n \end{Bmatrix}, \beta = \begin{Bmatrix} \beta_1 \\ \beta_2 \\ \cdots \\ \beta_n \end{Bmatrix}, X = \begin{bmatrix} 1 & x_{11} & x_{12} & \cdots & x_{1p} \\ 1 & x_{21} & x_{12} & \cdots & x_{2p} \\ \cdots & \cdots & \cdots & \cdots & \cdots \\ 1 & x_{n1} & x_{n2} & \cdots & x_{np} \end{bmatrix}, \varepsilon = \begin{Bmatrix} \varepsilon_1 \\ \varepsilon_2 \\ \cdots \\ \varepsilon_n \end{Bmatrix} \tag{2}$$

The regression model is expressed as (3).

$$T_1 = X\beta + \varepsilon, \varepsilon \sim N\left(0, \sigma^2 l_n\right) \tag{3}$$

where 0 is the n-dimensional 0 vector, the identity matrix of order n.

T2 prediction model of traffic accident duration. Under the condition of traffic accident, the deterministic queuing model can accurately express the accumulative arrival and departure states of vehicles at each stage. Based on the deterministic queuing model theory and the predicted value of T_1 established earlier can be used to calculate the recovery time T_2 of traffic events. Where, represents the time when a traffic incident occurs; $t_0 t_a$ Represents the end time of traffic incident duration; t_b Represents the end of traffic recovery time;T_1 Represents the duration of the event;T_2 Represents the traffic recovery time;C_0 Represents the lane capacity after the event is handled;C_1 Represents the capacity of the remaining lanes after the incident; Q

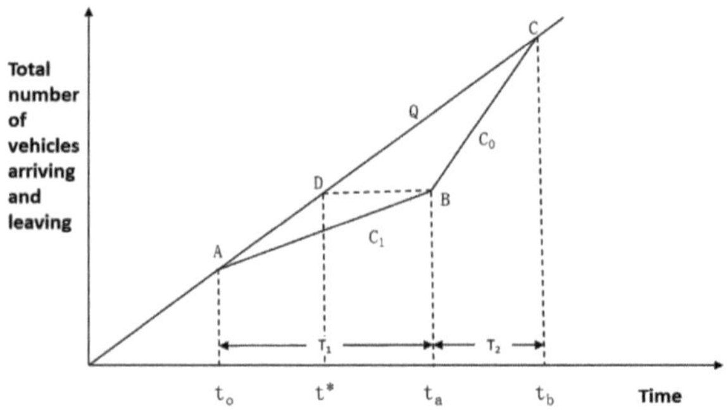

Fig. 2 Cumulative arrival and departure curves of vehicles

represents the upstream arrival rate and the slope of the cumulative arrival vehicle curve (Fig. 2).

Suppose Q and C_0, C_1 is constant. The value of C_0 can be determined according to the traffic flow parameters detected by the traffic detector, and then combined with the effective capacity coefficient table· We can determine the value of C_1.

The calculation formula of traffic recovery time t_b is as follows:

$$t_b = \frac{(C_0 - C_1)T_1}{C_0 - Q} + t_0 \tag{4}$$

The calculation formula of traffic recovery time T_2 is as follows:

$$T_2 = t_b - t_0 - T_1 = \frac{(C_0 - C_1)T_1}{C_0 - Q} \tag{5}$$

The formula for calculating the total duration of traffic incidents is:

$$T_A = T_1 + T_2 = \frac{(C_0 - C_1)T_1}{C_0 - Q} \tag{6}$$

3.2 *Traffic Event Queue Length Prediction*

When a traffic accident occurs on an expressway, its capacity will be affected. At this time, a gathering wave with uniform velocity propagating upstream will be generated at the event section, and its velocity is W_{01}; After a period, the traffic police will close part of the lane to deal with the accident site. At this moment, a new gathering wave will be generated with a speed of W_{12} And when the assembled wave W_{01} And dissipation wave W_{12} When meeting, the maximum queue length L_m will be generated at this moment. When the traffic accident is handled, the main line will restore the maximum capacity, and at this moment, the dissipating wave W_{23}

Table 1 Calculation model of traffic incident influence length

Number	Maximum influence length of mainline	Maximum influence length of connecting road
1	$L_m = \mathrm{W}_{02}\mathrm{T}_{02} + \mathrm{W}_{12}\mathrm{T}_{12}$	–
2	$L_m = L + W'_{23}T'_{23}$	$l_m = w_{02}t_{02} - l$
3	$L_m = L + W'_{23}T'_{23}$	$l_m = w_{12}t_{12} + w_{02}t_{02} - l$ or $l_m = w_{23}t_{23} + w_{13}t_{13} - l$
4	$L_m = \mathrm{W}_{23}\mathrm{T}_{23} + \mathrm{W}_{13}\mathrm{T}_{13}$	–
5	$L_m = L + W'_{13}T'_{13}$	$l_m = w_{13}t_{13} - l$
6	$L_m = L + W'_{23}T'_{23} + W'_{13}T'_{13}$	$l_m = w_{12}t_{12} + w_{02}t_{02} - lor$ $l_m = w_{23}t_{23} + w_{13}t_{13} - l$

spreading uniformly to the upstream section of the accident will be generated; In addition, the impact of ramps and connecting roads on traffic capacity should also be considered in the whole highway network environment. Due to the change of ramps or connecting roads into vehicles, a new aggregation wave W_{ij} will be generated.

The traffic wave theory is widely used in the analysis of the characteristics and influence of traffic flow fluctuation on expressways. The basic model is $W = \frac{\Delta q}{\Delta k}$, wherein Δ Q and Δ K represent the changes of traffic volume and traffic density before and after traffic accidents respectively, and W is the wave speed of traffic wave. Greenhill model is a speed-density linear model. Its basic model is $Q = KV_f\left(1 - \frac{K}{K_J}\right)$, where K represents density, V_f represents free flow velocity, and Q represents traffic volume. Based on the traffic wave theory and Greenhill model, this paper constructs the traffic event model of ramps and connecting roads to predict the queue length. The queue length of traffic events can be divided into the following six situations (Table 1):

3.3 Division of Inducing Information Release Area and Determination of Traffic Information Priority

In this study, the inducing information releasing area is divided into strong information releasing area and weak information releasing area. Travelers in the releasing area can decide whether to detour according to the congestion time and scope of traffic events. In reference [5], the release principles of strong prompt information and weak prompt information are further divided according to the distance from the traffic incident point. Strong prompt information has a higher priority than weak prompt information [5].

Among them, the area of strong prompt information release is determined based on the prediction model of the influence area of the traffic accident mentioned above, and the strong prompt information such as notice the accident ahead, reduce the speed, pay attention to avoid traffic and drive carefully is issued to the vehicles within this area. The weak warning information release area is determined according to the information effectiveness attenuation theory. Information validity can be specified by $\emptyset_{rs} = e^{kl}$, where $\emptyset_{rs}$ represents the validity of the information, k represents the correlation coefficient, which should be based on the results of the driver's questionnaire, and l represents the distance from the vehicle to the specified section. The farther away the operator and passenger is from the event point, the less effective the information is. The weak warning information release area is to remind vehicles within this area that they are about to enter the accident affected area and drive carefully.

3.4 Induced Information Advertising Policy

Through the establishment of the prediction model of the total duration of the accident and the model of the influence range of the traffic event, this paper further divides the release area of the traffic guidance information and determines the release range of the strong prompt information and the weak prompt information based on the priority of the information release. Then combined with the meteorological information, traffic event information, traffic network operation state and other information collected by the traffic detector, data analysis and further strategy generation are carried out. The adjusted spatio-temporal guidance information is published through VMS, mobile phone APP, vehicle-mounted terminal, roadside base station, traffic broadcast, Internet, and other terminals [6].

4 Experimental Verification

In this paper, a traffic accident in a highway in Shandong Province is selected. The traffic information release strategy is based on the premise that traffic congestion will occur when the traffic event occurs, and the case verification and analysis of the traffic guidance information release strategy are carried out. The data information can be retrieved from the database. The relevant information is shown in the following Table 2.

4.1 Prediction of Total Duration of Traffic Incidents

A total of 6,264 complete traffic event data were selected from a highway in Shandong province from 2019 to 2021, 56 traffic events with incorrect data were removed, and a total of 6,208 traffic accidents with valid data were removed. The duration of traffic accidents was defined as a dependent variable, and each factor affecting the duration of traffic incidents was defined as an independent variable. This paper

Table 2 Basic information

Attribute	Value	Attribute	Value
In the date	2021/2/8	The event type	Traffic accident (rear-end collision)
Time of occurrence	6:31 AM	Take up the driveway	2
Processing end time	8:46 AM	Number of vehicles involved	3
Processing duration	135 min	Length of congestion	30 km
Recovery time	10:42 AM	Number of goods vehicles involved	3
Event description	Three freight cars rear-ended each other	Number of truck rollovers	0

conducts variance analysis and regression analysis based on SPSS (Statistical Product and Service Solutions) simulation software. The significant factors influencing the duration of traffic events are determined by an ANOVA (Analysis of Variance) as shown in the following Table 3:

Based on the results of analysis of variance, independent variables with significance less than or equal to 0.04 are screened, including traffic accident type, discovery period, number of trucks involved, number of buses involved and number of congested lanes. Further regression analysis is carried out based on the screened significant factors, and non-standardized coefficients β (Mentioned in 3.1) are obtained as shown in the Table 4.

According to the non-standardized coefficients β of the respective variables, the optimal linear regression equation of the traffic accident duration in Eq. (2) is:

$$T_1 = 16.461 + 4.935X_1 + 12.116X_2 + 22.270X_3 + 7.84X_4 + 8.95X_5$$

Table 3 Intersubjective effect test

Source	Significance
Modified model	0.000
Intercept	0.000
Type of traffic accident	0.000
Find time	0.000
Number of trucks	0.000
The tanker number	0.053
Bus number	0.036
Number of vehicles involved	0.042
Number of blocked lanes	0.000

Table 4 Non-standardized coefficients β

Model	non-standardized coefficients β
Constant	16.461
Type of traffic accident	4.935
Find time	12.116
Number of trucks	22.270
Number of vehicles involved	7.840
Number of blocked lanes	8.952

Further, substitute the traffic accident case data selected in this paper, and get $T_1 = 2.098$ h.

Similarly, the traffic accident case data selected in this paper are substituted into Eqs. (5) and (6): $T_2 = 115.42$ min, $T_A = 4.02$ h.

Thus, the predicted value of the total duration of the traffic event is, where the predicted value of the traffic event duration is, and the predicted value of the traffic recovery time is. The accident is expected to end at 8:37am and recover at 10:32am.

4.2 *Prediction of Traffic Accident Influence Area and Determination of Inducing Information Release Area*

Based on the Greenshield model, the traffic volume Q and traffic density K values at each stage are shown in the following Table 5.

Based on the classical traffic wave theory, there is.

$W_{01} = \frac{Q_0-Q_1}{K_1-K_2} = 13.65; W_{12} = \frac{Q_1-Q_2}{K_1-K_2} = 147.47; W_{23} = \frac{Q_2-Q_0}{K_2-K_0} = 18.45;$

The traffic wave speed calculation data of traffic accident duration prediction above from Table 1 are substituted respectively to obtain $L_m = 31.55$ km.

4.3 *Comparative Analysis of Results*

The verification results are compared with the real data of a highway rear-end collision in Shandong selected in this paper, and the results are as follows (Table 6):

Table 5 Traffic volume and traffic density calculation

i	$Q_i(pcu/h)$	A formula to calculate	$K_i(pcu/km)$
1	$Q_0 = 3000$	$K = Q/v = 3000/100$	$K_0 = 30$
2	$Q_1 = C_1 = 1716$	$Q_1 = 100K\left(1 - \frac{K}{144}\right)$	$K_1 = 124.1$
3	$Q_2 = 4400$	$Q_2 = 100K\left(1 - \frac{K}{144}\right)$	$K_2 = 105.9$

Table 6 Data comparison

Compare the parameters	Actual accident data	Accident prediction data
Traffic accident handling time	135 min	125.85 min
Traffic accident recovery time	116 min	115.42 min
Maximum queue length of traffic accident	28 km	31.55 km

By comparing and analyzing the actual and predicted data of the traffic accident, it is found that the actual processing time, recovery time and the maximum queue length of the traffic accident are basically consistent with the predicted data, which can effectively predict the spatio-temporal range of the traffic accident.

5 Conclusions

Based on summarizing the inducing information publishing strategies at home and abroad, this paper firstly proposes a generalized information service system architecture which can be published in time through coordinated and efficient multi-channels, and further studies the inducing information spatio-temporal publishing strategies for generalized information services based on the traffic wave theory. The duration of traffic events and the influence range prediction model of traffic events are studied and established to determine the release time and space of highway traffic guidance information under traffic events, and then determine the time and space release strategy of highway traffic guidance information under traffic events. Finally, an example is analyzed and verified. Based on the analysis results can be found that this strategy can be to traffic managers to formulate scientific and reasonable traffic induced information release strategy of time and space to provide technical support, for users provide more scientific and accurate travel induced traffic information service, and further enhance the highway public travel information service level.

References

1. Wang, Z.W.: Research on information linkage release scheme under the events related to The Second Qin Expressway. Zhangjiakou Management Office of Erqin Expressway 7 2019
2. Zhang, X.T.: Study on traffic diversion method of expressway network. Souast University 5 (2018)
3. Xu, T.D., Hao, Y., Xu, X.H., Li X., Gao, X., Ma, L.: Regional group traffic guidance system and method considering driver dynamic response behavior. Shanghai Maritime University 5 2019

4. Guo, Y.R.: Research on short-term prediction and Guidance of traffic Flow based on Urban Road Traffic Data. Lanzhou University of Technology 1 2021
5. Chen, F., Zhang, W.H., Ding, H., Yan, P.: VMS induction strategy based on travelers' path choice behavior. School of Automotive and Traffic Engineering, Hefei University of Technology 5 (2018)
6. Yang, D.: Research on joint Publishing strategy of Variable information signs on expressways. Southeast University 3 (2016)

Research on Power IoT Intrusion Detection Method Based on Federated Learning

Guo Xiaoyan

Abstract The development of electricity Internet of Things (IoT) is potentially accompanied by an increase in network attack invasion. With the rapid development of artificial intelligence, machine learning, deep learning and other technologies are gradually applied to intrusion detection. However, the existing artificial intelligence program is seriously dependent on data, and with the attention of the world's attention to data privacy protection, the traditional artificial intelligence algorithm causes the model effect to be unsatisfactory because it cannot guarantee a certain amount of training. In order to protect the network security of the power network, this paper proposes an intrusion detection model based on federal learning, and uses a cluster algorithm to physically deploy an edge server, and use logic regression algorithm to network data. Intrusion detection. The intrusion detection model proposed in this paper not only protects the privacy of the grid user, but also determines the accuracy and robustness of the model on the data volume due to the problem of "data island" problem.

1 Introduction

The Intrusion Detection System (IDS) is a sense of malicious users by detecting and analyzing network traffic. At present, numerous artificial intelligence algorithms are applied in the intrusion detection system, and the algorithm is in theory, but it does not consider data issues when landing. Artificial intelligence algorithms rely on data for model training. In theory, the more data, the better the model effect.

Currently, in addition to a limited number of fields, more areas are just small data, or data from poor quality, and these data are also distributed in different mechanisms.

G. Xiaoyan (✉)
State Grid Tianjin Information & Telecommunication Company Key Laboratory of Energy Big Data Simulation of Tianjin Enterprise, Tianjin, China
e-mail: 13920604365@163.com

Information and Communication Company, State Grid Tianjin Electric Power Company, Hebei District, 153 Kunwei Road, Tianjin, China

L. C. Jain et al. (eds.), *Advances in Wireless Communications and Applications*,
Smart Innovation, Systems and Technologies 299,
https://doi.org/10.1007/978-981-19-2255-8_18

In December 2016, The National Power Grid launched an application for "electric e treasure", "handle-on power", and other applications. These APPs have disclosed, involving more than 10 million, and the hazard continues to expand. The EU has introduced the first name called 《General Data Protection Regulation》 Data Privacy Protection Act. It can be seen that the company collects, shares, and analyzes data in the case of user unaware. Just 20 companies and other more than 20 enterprises such as Tencent, Huawei committed "Do not listen to personal privacy". Traditional smart grid intrusion detection models are difficult to gather a lot of data for model training to ensure model effects.

This paper proposes a smart grid intrusion detection model based on federal learning. First, clustering clients are used to cluster, and the central deployment edge server of the class is used to solve the problem of limited client computing resources. Intrusion detection. Experimental tests were performed by the IDS2017 data set, and the experiment showed that the accuracy is fitted to the traditional centralized logic regression algorithm under the condition of protecting data privacy.

2 Research Status

At present, there are three main types of intrusion detection methods at home and abroad: are based on behavioral abnormal detection methods, based on rule-based misuse detection methods, and mixed detection methods [1]. The main advantage of behavioral abnormal detection methods is very sensitive to new attack types, which is conducive to detection of new attacks, and can detect zero-day vulnerabilities; disadvantages are high training costs, complex learning behavior, and easy to produce higher false packets. Rule-based misuse detection method is to extract feature data in a large number of attack data, which will meet the data of the matching rule as an attack behavior, but the maintenance cost is relatively high, and the new attack type is not sensitive, can be carried by attack load Processing bypassing the rules match. The mixed intrusion detection method is to combine the first two types to improve the detection rate of the known intrusion type and reduce the false positive rate of future attack types.

In [2], Ren proposed a structure of a fusion zone chain and a federal learning, designed an intrusion detection algorithm for lightweight network equipment. Document [3] proposes an intrusion detection method based on a KPCA method and Support Vector Machine algorithm (SVM). Document [4] proposes an intrusion model based on internal and external convolution networks, and the accuracy and false positive rate of the modeling model are better than the baseline model. Document [5] uses depth migration technology to provide an intrusion detection algorithm for the Internet of Things, mainly identifying distributed denial services, data mobile phones, etc., and the proposed method has high classification accuracy. Literature [6] Chen improve the traditional use of signal dual profiles to improve, using the Support Vector Machine to invade the wireless device terminal, the method has a good accuracy.

Today, machine learning technology is widely used in intrusion detection systems. This paper utilizes a classification algorithm logistic regression to training sets, and obtains weights of each network eigenvalue, which predicts results.

3 Related Basics

3.1 Federated Learning

Traditional artificial intelligence algorithms require users to train data to the center server. However, the collected data may contain many privacy information, and the user may not be willing to disclose private information, or the user refuses to share data. Second, the existence of the Internet technology giants monopolizes a lot of data. This is also the cause of "data island", causing difficult to share between the data, it is difficult to create the data value of "1 + 1 > 2". Today, even the data sharing channel between the different departments of the same company is not easy to open. At the same time, the global scope has brought more challenges to data sharing on data privacy and security. The EU introduces 《General Data Protection Regulation 》GDPR, any organization that involves personal information will be bound by it. Facebook and Google have become the first argument after this bill takes effect.

In 2016,Google takes the lead in proposing the federal learning privacy protection framework to use mobile phone users' local keyboard input data to predict the next prompt input of the user's keyboard. A shared model is established between the server and the mobile terminal, so that the data "can be invisible" can be implemented in the case of mobile terminal data, and the user privacy has also reached the purpose of data utilization.

In this frame, the server first initializes the information of the model, each distributed terminal downloads the initialization model in the server, and then each distributed terminal is based on the local data set, and only the gradient information of the model is transmitted after training. The server receives the gradient information of each distributed terminal to integrate according to the proportion of the data set of each terminal. The above steps are a round of training, the training ends when the number of preset iterations or the satisfactory effect is achieved.

Suppose there are N terminals in the federal learning system, each terminal has $n_i(1 \leq i \leq N)$ sample, then the loss function of the central server is as in Eq. (1):

$$f_j(w) = \sum_{i=1}^{N} \frac{n_i}{n} F_i(w) \tag{1}$$

where $F_i(w) = \frac{1}{n_i} \sum_{j=1}^{n_i} f_j(w)$ is the loss function of the i-th terminal, $f_j(w)$ is the j-th sample of the i-th terminal Loss, w is the model parameters of the current echo.

Federal learning generally adopts a gradient decrease algorithm to minimize the loss function until the number of specified iterations or a certain model accuracy is reached.

3.2 Logistic Regression Classification Algorithm

The hypothesis function of the logistic regression [8, 9] is used to use the hypothesis function of the linear regression, so the weight vector w obtained in the former can be well shown in an impact of a feature vector to the classification result.

The hypothesis function of logistic regression is as in Eq. (2).($\boldsymbol{x}$ is the feature vector, $\boldsymbol{w}$ is the weight vector).

$$\boldsymbol{h}_{\boldsymbol{w}}(\boldsymbol{x}) = \frac{1}{1 + e^{-W_x^T}} \tag{2}$$

Loss function, such as Eq. (3) ($\boldsymbol{y}$ is the true label value, $\boldsymbol{h}$ is the predicted label value, $\boldsymbol{m}$ represents the total number of data).

$$F(w) = -\frac{1}{m}\sum_{i=1}^{m}\left[y_i * \log(h_w(x_i)) + (1 - y_i) * (1 - h_w(x_i))\right] \tag{3}$$

The gradient decrease algorithm finds a certain set of weights $\boldsymbol{w}$ make $\boldsymbol{F}(\boldsymbol{w})$ as small as possible, as in Eq. (4) ($\boldsymbol{\alpha}$ is step size).

$$w_i^j = w_i^{j-1} - a * \frac{\partial F}{\partial w_i} \tag{4}$$

4 Power IoT Intrusion Detection Algorithm Based on Federal Learning

4.1 Power IoT Intrusion Detection Algorithm Framework

The construct of intrusion detection system requires a lot of effective label data to analyze learning. For small and medium power grid units, the network data that can be collected is very limited, and tags are also required mark. This paper uses the federal machine learning framework to achieve a distributed machine learning task for more power grid units, solve the traditional "data island" problem [7], aggregate more available data. In federal learning, task participants do not need to share local

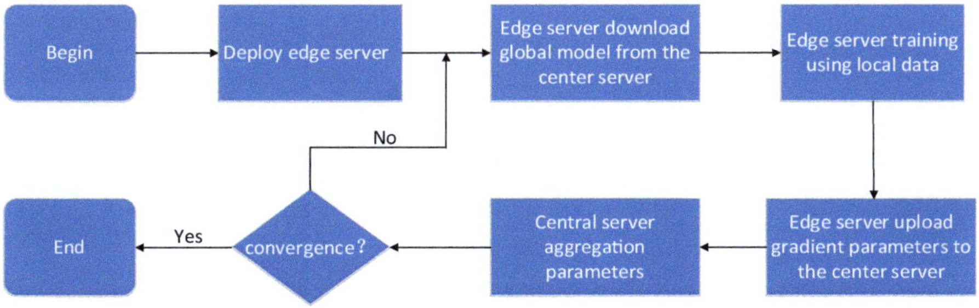

Fig. 1 Intrusion detection framework process scheme

data, and only need to train, only need to interact and update the iterative gradient parameters. The workflow of this framework is shown in Fig. 1.

First, in order to ensure sufficient computing power and stable communication capabilities, we deploy edge servers based on physical distance through a clustering algorithm. Secondly, the edge server collects the raw data of the power grid belonging to each terminal within its own range. Finally, the model training of the logistic regression algorithm under federated learning. The edge server uses the model aggregated by the central server in the previous round to perform a new round of model training, and then uploads local model parameters. Repeat this until convergence, and its simple process is shown in Fig. 1. The logistic regression model trained with power grid data can perform well intrusion detection and discrimination. Because it aggregates grid data of a large number of distributed nodes, theoretically the more data, the better the effect and robustness of the model.

The intrusion detection framework based on federated learning proposed in this paper mainly includes three main characters of central server, edge servers, and power users. Their respective functions are as follows:

Power User: The actual owner of network data, that is the user in the smart grid, the smart meter is responsible for collecting the user's information to the edge server.

Edge Server: Small servers deployed within a certain physical link range is to solve user-end computing, storage resource limited problems, responsible for collecting scope users to model training, and interact with the cloud server. User data does not have an edge server associated with it to a certain extent, to a certain extent, to protect data privacy.

Central Server: The role of publishing initial model, collection, integration, distribution global model, constitutes server cluster, preventing single point fault issues, ensuring normal implementation of training tasks.

The framework structure mentioned in this paper is shown in Fig. 2, mainly divided into cloud, side, and households. The cloud is a central server working layer, and the edge layer is deployed with an edge server. The user layer is the actual owner and producer of the data.

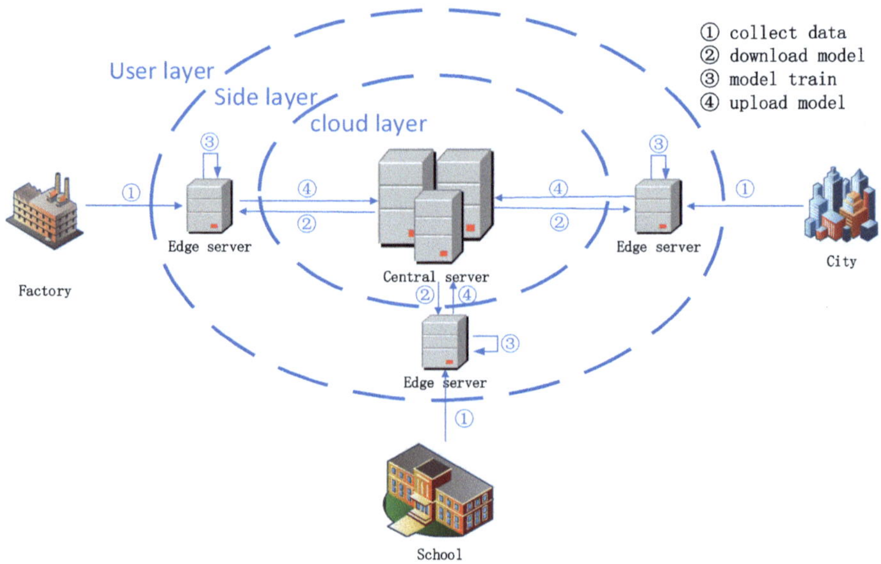

Fig. 2 Power IoT intrusion detect framework

4.2 Intrusion Detection Algorithm Experiment

Environment: cpuR7-5800H, gpu3060, Python 3.9,Windows 11.

The data set of this experiment is IDS2017, about 2,830,700 data, 81 features (including Flow Duration, Total Forward Packets, Total Backward Packets,Total Length of Forward Packets, Total Length of Backward Packets,, etc.). Randomly disrupt 2,830,700 data, extract data volume ratio of 0.01, totaling 28,307 data, and divided into training set and test set by pressing 3: 1 ratio. The data label has 11 attack prediction results such as Benign, DOS Hulk, DDoS, BOT, respectively correspond to 0, 1, …, 11, respectively.

Experiment 1: Simulate three edge nodes (three institutions or companies in reality), divide the 2,830,700 * 0.75 data is divided into 3: 3: 4, and finally according to the federal learning method data on three edge nodes Through joint training, the change in accuracy with the number of times is shown in Fig. 3.

Experiment 2: Simulate five edge nodes (five institutions or companies in reality), the proportion of 2,830,700 * 0.75 data is 2: 2: 2: 2: 2: 2: 2: 2: Finally according to federal learning methods The data of the edge node performs joint training, the change in the number of times the number of times is shown in Fig. 4.

The experimental results show that compared with the traditional single-node logic regression algorithm based on federal learning, the power-based logic regression algorithm is not only protected by user data privacy, but also accurate comparison. The accuracy of some separate training of some nodes is higher. To a certain extent, because of the “Data Lone Island”, many data have been effectively utilized, making the robustness of the model better.

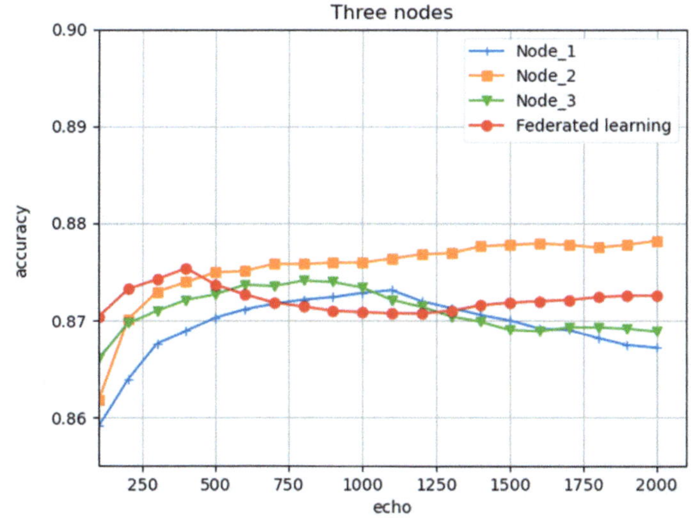

Fig. 3 Detection accuracy of three edge nodes

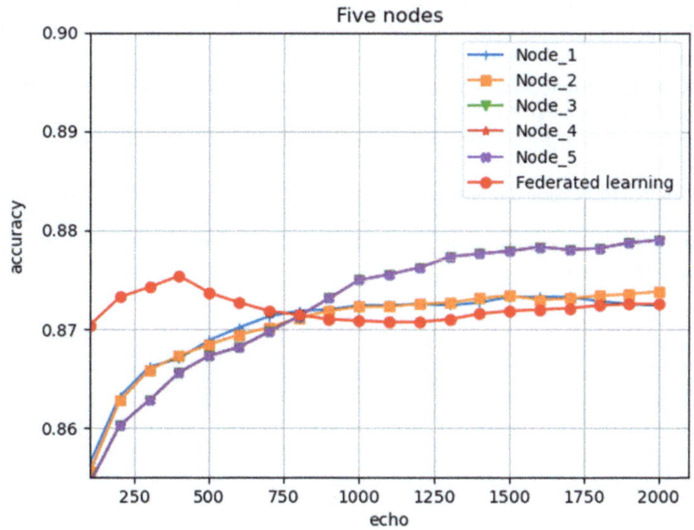

Fig. 4 Detection accuracy of five edge nodes

5 Conclusion

This paper proposes a power IoT intrusion detection method based federal learning. Federal learning as a revolutionary innovative privacy protection new paradigm, which can protect data security while user data, while protecting data security, is a unified modeling training for data information from various nodes. This paper is tested on the IDS2017 data set to achieve the expected effect. Comparing data centralized training methods, this paper does not sacrifice larger precision to achieve distributed training purposes, protecting data privacy, breaking "Data Lone Island" to a certain extent, increasing the robustness of the model Prevent it from fitting. Future work takes into account the distributed characteristics of federal learning and block chains, enabling distributed IDS, improve electricity network security, and protects residents.

References

1. Sun, Y.Z.: Application of artificial intelligence in electricity network intrusion detection. China Inform. Security **06**, 45–47 (2021)
2. Ren, T.: Network intrusion detection algorithm for fusion zone chain and federal learning. Inform. Network Security **21**(07), 27–34 (2021)
3. Shone, N., Ngoc, T.N., Phai, V.D.: A deep learning approach to network intrusion detection. IEEE Trans. Emerg. Top Computlntell. **2**(2), 41–50 (2018)
4. Wang, Y.F.: Network intrusion detection based on internal and external convolutional network. J. Beijing University of Posts Telecommunications, 1–7 (2021)
5. Zhan, L.J.: Internet invasion detection framework based on deep migration learning. Internet of Things Tech. **11**(11), 58–61 (2021)
6. Chen, Z.X.: Intrusion detection algorithm based on grid wireless device based on radio frequency fingerprint. Radio Eng. **51**(05), 352–359 (2021)
7. Yang, Q.: AI and data privacy protection: Federal learning crack. Information Security Res. **5**(11), 961–965 (2019)
8. Mao, Y.: Logical regression model based on density estimation. Automatic Chem. **40**(01), 62–72 (2014)
9. Guo, H.P.: Logical regression method for class imbalance. Pattern Identifi. Artificial Intell. **28**(08), 686–693 (2015)

Traffic Flow Prediction of Expressway Section Based on RBF Neural Network Model

Qun Liu, Zhuocheng Yang, and Lei Cai

Abstract In order to further improve the prediction accuracy of expressway traffic flow, this study proposed an RBF neural network model. Firstly, RBF is used to train the model by using ETC (Electronic Toll Collection) gantry historical data considering the time-varying characteristics of the flow, to ensure the similarity of the flow curve and robustness of the model. Then, taking three typical ETC gantries from thirty gantries of Beijing-Shanghai Expressway in Shandong province as an example, the accuracy of the model is verified by using the historical operation data of them during holidays. The results show that: (1) The flow of ETC gantry section in holidays predicted by RBF is closer to the actual value, and the prediction accuracy is significantly better than that of BP and ELMAN. (2) The MAE is within 75 veh/min, the RMSE is within 6veh/min, and the MAPE is less than 4.5%.

1 Introduction

ETC gantry data are traffic flow section data, accurately record different types of vehicles and key traffic characteristics such as speed and flow. The accuracy of data is more than 99%, compared to other data sources such as traffic survey data, RTMS (Remote Traffic Microwave Sensor) data. The integrity, accuracy and authenticity of ETC gantry data are better than them. The data can help managers and decision makers to make more effective traffic flow prediction.

In the field of traffic flow prediction, scholars have been proposed parameter models, nonparametric models, hybrid models and machine learning/deep learning models [1]. Neural network models are a subclass of nonparametric model, which have widely interconnected structure and effective learning mechanism to simulate

Q. Liu
Shandong High-Speed Construction Management Group Co., Ltd, Jinan 250101, China

Z. Yang (✉) · L. Cai
Beijing GOTEC ITS Technology Co., Ltd, Beijing 100088, China
e-mail: yzc@itsc.cn

L. C. Jain et al. (eds.), *Advances in Wireless Communications and Applications*,
Smart Innovation, Systems and Technologies 299,
https://doi.org/10.1007/978-981-19-2255-8_19

the process of the human brain information processing. Based on a large amount of historical data for training, high prediction accuracy is achieved [2–5].

In order to improve the prediction accuracy and make full use of ETC gantry data information, this study used RBF neural network model to predict traffic flow based on historical data. At the same time, the accuracy of the prediction curve is guaranteed from the time-varying characteristics of the flow. The proposed method has high prediction accuracy and strong robustness, and provides an idea for the prediction of section traffic flow.

2 ETC Gantry System

In 2019, China vigorously promoted ETC technology on 143,000 km of expressways, cancelled 487 toll stations at provincial boundaries of expressways, built 24,588 sets of ETC gantry systems, reconstructed 48,211 ETC lanes. The number of ETC users reached 2.0400 million. Highway infrastructure and management level to achieve qualitative progress by leaps and bounds, obtain a series of surprising results.

According to the overall technical requirements of the expressway provincial boundary toll stations, ETC gantry systems should be set up between each interchange and entrance/exit of expressways. ETC vehicles and MTC (Manual Toll Collection) vehicles realized segmented tolling. Generating transaction flow (or pass certificate), ETC pass record and captured image information (including license plate number and license plate color, etc.) for ETC vehicles, and timely upload to provincial settlement center and ministry network center. For MTC vehicles, read vehicle information in CPC card (including license plate number, license plate color, model information, etc.), calculate the fee and write it into CPC card, form CPC line record, and upload the captured image information to provincial settlement center and ministry network center in time.

Through the detailed study on the content of the data of ETC gantry systems, it can be found that ETC gantry systems can obtain high-precision information such as traffic flow and interval average speed, and can evaluate the real-time traffic running state of expressways. Identify the traffic congestion near gantry or predict the coming traffic congestion is of great important.

3 Overview of RBF Neural Network

RBF (Radial Basis Function) neural network is a typical feedforward neural network. Its characteristic is that the radial basis function is used as the transformation function of the nodes in the hidden layer, so that the hidden layer can convert the low-dimensional input data into the high-dimensional space, and convert the linear non-separable problem in the low-dimensional space into the linearly separable problem in the high-dimensional space. RBF neural network is usually composed of input layer, hidden layer and output layer, and its structure is shown in the Fig. 1 [6, 7].

Vector $X(X = (x_t, x_{t-1}, \ldots, x_{t-n}))$ is the input variable of the network, vector $Y(Y = (\hat{x}_{t+1}, \ldots, \hat{x}_{t+d}))$ is the output variable of the network, vector W represents the connection weight matrix between the input layer and the hidden layer,.

The prediction process of RBF neural network can be expressed as:

$$w_j = \exp\left(-\frac{x_{t-j} - u_j^2}{2\sigma_j^2}\right) \tag{1}$$

$$\hat{x}_{t+d} = \sum w_{jd} w_j \tag{2}$$

In this formula, w_j is the output of the jth node in the hidden layer, x_{t-j} is the traffic flow observation value at $t - j$ moment, $\|x_{t-j} - u_j\|$ is the normal function, u_j is the center of the Gaussian function, and σ_j is the variance of the Gaussian function. n indicates the number of nodes of the input layer, m indicates the number of nodes of the hidden layer, d indicates the number of nodes of the output layer, w_{jd} indicates the connection weight between the output layer and the hidden layer, and $\hat{x}_{t+d}$ indicates the predicted traffic flow at $t + d$.

The center vector of the radial basis function $u_j = \left[u_{j1}, u_{j2}, \ldots, u_{j(t-n)}\right]^T$. The kernel width σ_j and the connection weights w_{jd} of hidden layer and output layer are parameters of RBF neural network. u_j and σ_j can be determined by FCM clustering algorithm in Eqs. (3) and (4), and the parameter w_{jd} is obtained by gradient descent learning algorithm.

$$u_{jk} = \sum_{i=1}^{n} \mu_{ij} x_{ik} / \sum_{i=1}^{n} \mu_{ij} \tag{3}$$

$$\sigma_j = \sum_{i=1}^{n} \mu_{ij} x_i - u_j^2 / \sum_{i=1}^{n} \mu_{ij} \tag{4}$$

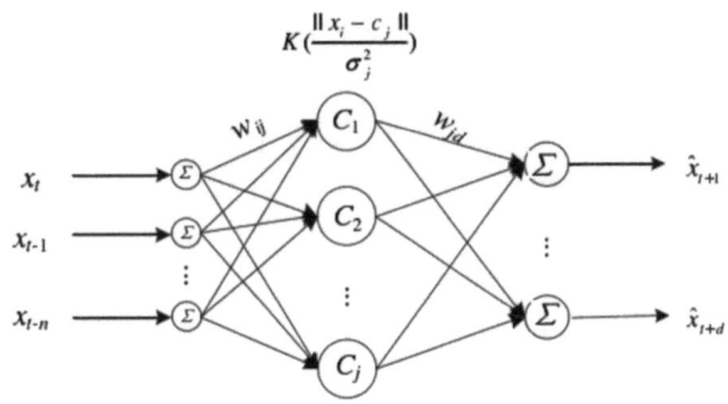

Fig. 1 Neural network prediction principle

In the formula, μ_{ij} represents the fuzzy membership degree of x_i of the sample obtained by FCM clustering algorithm for the jth class, and n represents the training sample size.

Let $\widetilde{x_j} = \varphi\|x_{t-j} - u_j\|$, $j = 1, 2, \ldots, m$, so

$$\tilde{x} = [\widetilde{x_1}, \widetilde{x_2} \ldots, \widetilde{x_m}]^T \tag{5}$$

The center u_j and the kernel width σ_j of the radial basis function obtained by Eq. (3) and (4) are substituted into Eq. (1) to realize the nonlinear mapping from the input layer to the hidden layer.

Then, start building the model as follows:

Step 1: According to formula (3) and (4), the values of u_j and σ_j are obtained, and the input model $\tilde{x}$ is also created according to formula (5).

Step 2: Introduce ε insensitive loss function.

ε insensitive loss function $L^{\varepsilon}(x, y, f)$ is defined as

$$L^{\varepsilon}(x, y, f) = |y - f(x)|_{\varepsilon} = \max(0, |y - f(x)|_{\varepsilon}) \tag{6}$$

In the formula, $x \in R^m$, $y \in R$.

For the linear model of formula (6), its corresponding ε insensitive loss function can be expressed as:

$$\sum_{j=1}^{n} \left|y_j^o - y_j\right|_{\varepsilon} = \sum_{j=1}^{n} \max\left(0, \left|y_j^o - y_j\right| - \varepsilon\right) = \sum_{j=1}^{n} \max\left(0, \left|p^T \widetilde{x_j} - y_j\right| - \varepsilon\right) \tag{7}$$

In the formula, y_j^o represents neural network output and y_j represents real output.

Step 3: Prediction.

$$y = p^T \varphi(\tilde{x}_{test}) = \lambda \sum_{j=1}^{n} \left(\alpha_j - \alpha_j^*\right) \varphi^T\left(\widetilde{x_j}\right)\left(\tilde{x}_{test}\right) = \lambda \sum_{j=1}^{n} \left(\alpha_j - \alpha_j^*\right) \tilde{K}\left(\widetilde{x_j}, \tilde{x}_{test}\right) \tag{8}$$

$$\begin{aligned} y &= [y_1 \ldots y_n]^T, \alpha = [\alpha_1 \ldots \alpha_n]^T, \alpha^* = \left[\alpha_1^* \ldots \alpha_n^*\right]^T, \\ \tilde{K} &= \left[\tilde{k}\left(\widetilde{x_j}, \widetilde{x_i}\right)\right] = \begin{bmatrix} K + \frac{\mu n}{\lambda} I & -K \\ -K & K \frac{\mu n}{\lambda} I \end{bmatrix} \end{aligned} \tag{9}$$

4 Experimental Verification

4.1 Basic Data Description

The data used in this study is the ETC gantry data of Shandong province. Firstly, the protocol and format of ETC gantry data are studied, and the data are initialized to determine the parameters of the model. On this basis, the prediction effect of the model was evaluated by root mean square error (RMSE), mean absolute percentage error (MAPE) and mean absolute error (MAE).

Basic data

In order to verify the validity and correctness of the model, three typical gantries of Beijing-Shanghai Expressway in Shandong Province were selected as the research objects, and the traffic flow during holidays was predicted. Among them, three gantry numbers are G000237011000320070, G000237011000420080 and G000237011000510040, all of which are uplink gantries. In this study, the historical gantry data of 8 days from September 29 to October 6, 2020 were selected to predict the traffic flow on October 7 and 8.

The ETC gantry data selected in this study includes ETC transaction flow (double-chip OBU), ETC traffic record (transaction failure), image flow record and CPC card record, among which the image flow record is the auxiliary data. Take ETC transaction flow (double-chip OBU) as an example, and its sample data are shown in Fig. 2.

ListNo	SerialNo	FlagNet...	FlagRoa...	FlagID	OBUType	OBUMa...	VehicleT...	VehPlate	VehColor	OpTime	VehStat...
G00023...	0137012...	G2	G00023...	G00023...	2	6867B296	13	冀D2E877	1	2020-10...	27
G00023...	0137012...	G2	G00023...	G00023...	2	6867B296	13	冀D2E877	1	2020-10...	27
G00023...	0137012...	G2	G00023...	G00023...	2	6867B2...	13	蒙D44335	1	2020-10...	27
G00023...	0137012...	G2	G00023...	G00023...	2	6867B2...	13	蒙D44335	1	2020-10...	27
G00023...	0137012...	G2	G00023...	G00023...	2	6867B2...	13	蒙D44335	1	2020-10...	27
G00023...	0137012...	G2	G00023...	G00023...	2	6867B2...	13	蒙D44335	1	2020-10...	27
G00023...	0137012...	G2	G00023...	G00023...	2	6867B2...	13	蒙D44335	1	2020-10...	27
G00023...	0137012...	G2	G00023...	G00023...	2	6867B2...	13	蒙D44335	1	2020-10...	27
G00023...	0137012...	G2	G00023...	G00023...	2	6867B2...	13	蒙D44335	1	2020-10...	27
G00023...	0137012...	G2	G00023...	G00023...	2	6867B2...	13	蒙D44335	1	2020-10...	27
G00023...	0137012...	G2	G00023...	G00023...	2	6867B2...	13	蒙D44335	1	2020-10...	27
G00023...	0137012...	G2	G00023...	G00023...	2	6867B2...	13	蒙D44335	1	2020-10...	27
G00023...	0137012...	G2	G00023...	G00023...	2	6867B2...	13	蒙D44335	1	2020-10...	27
G00023...	0137012...	G2	G00023...	G00023...	2	6867B2...	13	鲁JC7076	1	2020-10...	27
G00023...	0137012...	G2	G00023...	G00023...	2	6867B2...	13	鲁JC7076	1	2020-10...	27
G00023...	0137012...	G2	G00023...	G00023...	2	6867B2...	13	鲁JC7076	1	2020-10...	27
G00023...	0137012...	G2	G00023...	G00023...	2	6867B2...	13	鲁JC7076	1	2020-10...	27
G00023...	0137012...	G2	G00023...	G00023...	2	6867B2...	13	鲁JC7076	1	2020-10...	27
G00023...	0137012...	G2	G00023...	G00023...	2	6867B2...	13	鲁JC7076	1	2020-10...	27
G00023...	0137012...	G2	G00023...	G00023...	2	6867B2...	13	鲁JC7076	1	2020-10...	27
G00023...	0137012...	G2	G00023...	G00023...	2	6867B2...	13	鲁JC7076	1	2020-10...	27
G00023...	0137012...	G2	G00023...	G00023...	2	6867B2...	13	鲁JC7076	1	2020-10...	27
G00023...	0137012...	G2	G00023...	G00023...	2	6867B2...	13	鲁JC7076	1	2020-10...	27
G00023...	0137012...	G2	G00023...	G00023...	2	6867B2...	13	鲁JD0562	1	2020-10...	27

Fig. 2 An example of ETC transaction flow (two-chip OBU) data

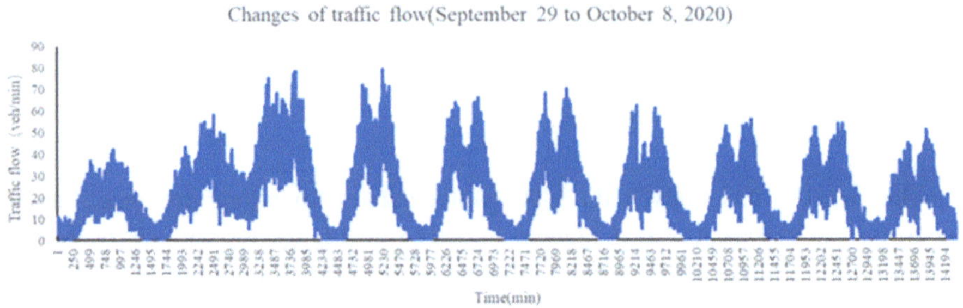

Fig. 3 Gantry G000237011000510040 traffic changes per minute during holidays

Data characteristics

The changes of traffic flow of gantry G000237011000510040 within the range affected by holidays in 10 days is shown in the Fig. 3. The average daily traffic flow value increases from September 29, and reaches its maximum value on October 2. It becomes stable on October 3 and 4, and gradually decreases on October 5, which conforms to the traffic flow rule of National Day holiday.

4.2 Prediction Results

The training process of all models was realized in MATLAB R2020b. The calculation formulas of the prediction result evaluation index are as follows:

$$MAE = \frac{1}{n}\sum_{i=1}^{n}\left|\hat{v}_i - v_i\right| \tag{10}$$

$$MAPE = \frac{1}{n}\sum_{i=1}^{n}\frac{\left|\hat{v}_i - v_i\right|}{v_i}*100\% \tag{11}$$

$$RMSE = \sqrt{\sum_{i=1}^{n}\frac{\left(\hat{v}_i - v_i\right)^2}{n}} \tag{12}$$

As shown in this formula, $\hat{v}_i$ indicates the predicted value of traffic flow at time i, and v_i indicates the observed value of traffic flow at time i.

Figure 4 shows a comparison of the traffic flow of three gantries on holidays by using BP, ELMAN and RBF algorithms. It can be seen from the figure that the prediction results of the three gantries obtained by RBF neural network algorithm are better than the other two methods, and the prediction errors are different due to the complexity of the sections where the three gantries belongs. The prediction

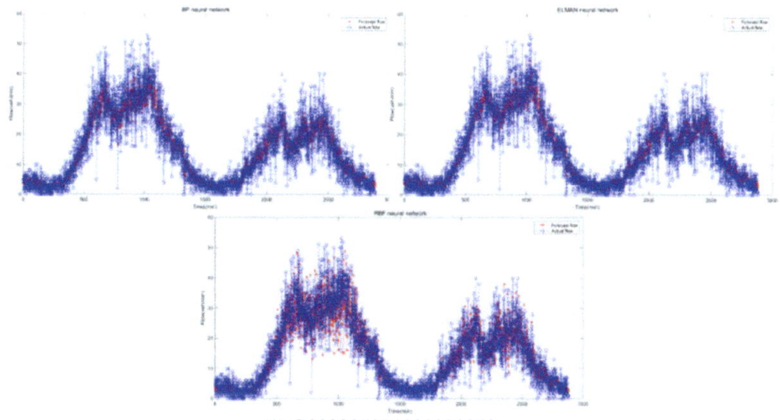

(1) G000237011000320070

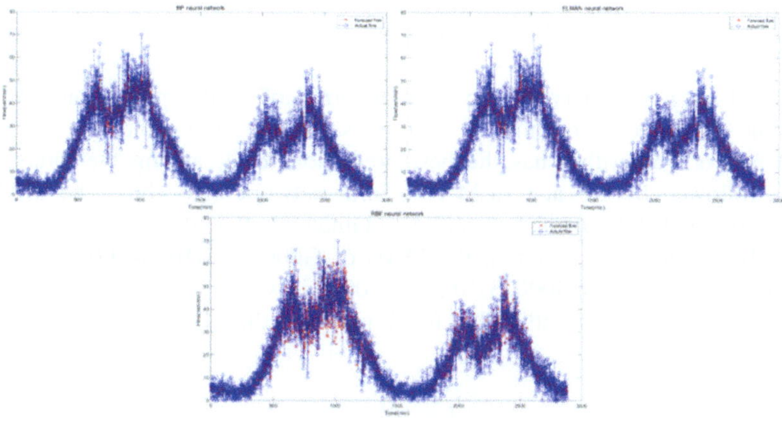

(2) G000237011000420080

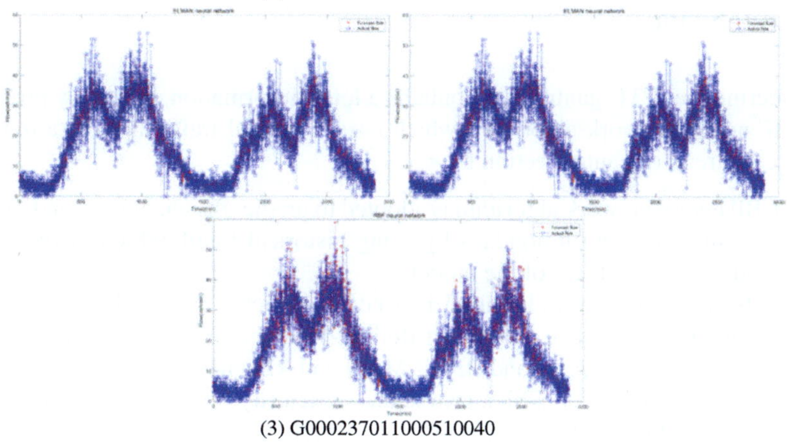

(3) G000237011000510040

Fig. 4 Comparison of holiday flow forecast results of BP, ELMAN and RBF

Table 1 Holiday traffic prediction error

The number of gantries	Algorithm	MAE/ (veh/min)	RMSE/ (veh/min)	MAPE/ (%)
G000237011000320070	BP	84.86	5.19	3.78
	ELMAN	83.06	5.12	3.74
	RBF	74.99	5.07	3.70
G000237011000420080	BP	59.87	5.98	4.41
	ELMAN	58.99	5.96	4.39
	RBF	54.14	5.91	4.36
G000237011000510040	BP	34.51	5.55	4.05
	ELMAN	33.72	5.53	4.04
	RBF	31.78	5.49	4.01

results of G000237011000510040 gantry are significantly better than the other two gantries.

Table 1 is a summary of the average error of traffic prediction obtained by using the three algorithms. It can be seen from the table that the RBF neural network algorithm has the best prediction result among the three algorithms, followed by BP and ELMAN neural network algorithm. The MAE of G000237011000320070 holiday traffic obtained by the algorithm proposed in this study is within 75 veh/min, the MAE of G000237011000420080 is within 55 veh/min, the MAE of G000237011000510040 is within 32 veh/min. The RMSE of the three gantries were all within 6 veh/min, and the MAPE were all less than 4.5%.

5 Conclusion

Considering that ETC gantry data contains a lot of information, this study proposes an RBF neural network algorithm, which uses historical traffic flow data trend to predict the trend of gantry section flow.

(1) RBF neural network algorithm is adopted to predict the changing trend of flow of gantry section on holidays by using historical trend, which increases the prediction robustness of the model.
(2) RBF neural network algorithm is used to predict the flow of gantry section during holidays: three gantries of Beijing-Shanghai Expressway in Shandong province were selected and BP, ELMAN and RBF neural network algorithms were used to predict, which proves the superiority of the proposed algorithm in this study. The MAE were less than 75veh/min, and the RMSE were less than 6veh/min, and the MAPE were less than 4.5%.

References

1. Liu, J., Guan, W.: A summary of traffic flow forecasting methods. J. Highw. Traffic Sci. Technol **03**, 82–85 (2004). (in Chinese)
2. Li, Y., Yu, R., Shahabi, C., et al.: Diffusion convolutional recurrent neural network: data-driven traffic forecasting (2017)
3. Zhang, J., Zheng, Y., Sun, J., et al.: Flow prediction in spatio-temporal networks based on multitask deep learning. IEEE Trans. Knowl. Data Eng. **1**(1) (2019)
4. Mou, L., Zhao, P., Xie, H., et al.: T-LSTM: a long short-term memory neural network enhanced by temporal information for traffic flow prediction. IEEE Access **7**, 98053–98060 (2019)
5. Zhao, L., Song, Y., Zhang, C., et al.: T-GCN: a temporal graph convolutional network for traffic prediction. IEEE Trans. Intell. Transp. Syst. **99**, 1–11 (2019)
6. Guo, S., Lin, Y., Feng, N., et al. Attention based spatial-temporal graph convolutional networks for traffic flow forecasting. In: Proceedings of the AAAI conference on artificial intelligence, vol. 33, pp. 922-929 (2019)
7. Song, C., Lin, Y., Guo, S., et al.: Spatial-temporal synchronous graph convolutional networks: a new framework for spatial-temporal network data forecasting. In: Proceedings of the AAAI conference on artificial intelligence, vol. 34(1), pp. 914–921, (2020)

An Adaptive Multi-component LFM Signal Parameter Estimation Based on STFRFT

Yu Dai, Xuan Wang, Wenjia Long, and Junyu Ma

Abstract In this paper, based on the multi-component linear frequency modulation (LFM) signal parameter estimation method, the fuselage echo model established and the rotor echo model within a short observation time are approximated to the characteristics of the LFM signal. Aiming at the problem of low accuracy of parameter estimation caused by the mutual influence of the component signals in the parameter estimation of multi-component LFM signals, an adaptive multi-component LFM based on short-time fractional Fourier transform (STFRFT) is proposed. This method combines the idea of parameter estimation from coarse to fine, adopts the strategy of reconstructing the signal and removing the estimated LFM component signal from the original signal, and realizes the parameter estimation of each LFM component signal in turn. According to the proposed presence or absence of the LFM component signal, the adaptive decision parameter estimation is terminated. The simulation experiment results verify the effectiveness of the method proposed in this paper.

1 Introduction

LFM signal has the characteristics of large time-wide bandwidth product. Using it as a transmission signal and adopting pulse compression technology can effectively solve the contradiction between detection range and range resolution, and is widely used in radar, communications, seismic exploration and other fields [1–3]. On the one hand, the target echo is in the form of a chirp signal without considering the micro-movement of the target rotating part. On the other hand, according to the previous analysis, the Doppler frequency of the strong scattering point of the rotor blade changes as a sinusoidal signal. In addition, in a short observation time, the echo signal whose Doppler frequency changes as a sinusoidal signal can be approximated as LFM signal [4]. Therefore, the estimation of the rotational angular velocity of the rotor blade can be transformed into the parameter estimation problem of the

Y. Dai · X. Wang (✉) · W. Long · J. Ma
Zhixing College of Hubei University, Wuhan, China
e-mail: winnie266@163.com

L. C. Jain et al. (eds.), *Advances in Wireless Communications and Applications*,
Smart Innovation, Systems and Technologies 299,
https://doi.org/10.1007/978-981-19-2255-8_20

LFM signal. From this point of view, accurate parameter estimation of LFM signals, especially multi-component LFM signals, is particularly important in the field of signal processing.

In recent years, many scholars have devoted themselves to improving the accuracy of PPS parameter estimation, including the STFT-based QML method proposed by Igor Djurović in 2014 [5], which uses STFT to estimate the instantaneous frequency of a polynomial phase signal, and adopts polynomial regression method to realize the parameter estimation of PPS. Compared with traditional methods such as HAF and PHAF [6, 7], although this method has a lower signal-to-noise ratio threshold and when the signal-to-noise ratio is high, the root-mean-square error value of the parameter estimation reaches Cramer-Rao lower bound [8], and LFM signal as a second-order simple polynomial phase signal, using this method can quickly and accurately achieve parameter estimation, but because STFT has a lower time–frequency resolution, it affects the estimation performance of the parameters and the noise ratio threshold, and this method can only be used to achieve single-component polynomial phase signals. In response to this problem, in 2017 Igor Djurović proposed a combination of STFT-based quasi-maximum likelihood estimation method and sequential elimination of estimated signals to achieve multi-component polynomial phase signal parameter estimation [9]. But this method is only limited, because the signal amplitude of each component differs greatly. When the signal amplitudes of different components are not much different, using STFT to estimate the instantaneous frequency of the component signal seriously deteriorates the estimation performance of the parameters. In addition, the method defaults that the number of component signals is known and no decision conditions for terminating signal parameter estimation are given.

This paper presents an adaptive multi-component LFM signal parameter estimation method based on STFRFT. This method uses STFRFT, which has a higher time–frequency resolution than STFT, to improve the estimation accuracy of the instantaneous frequency of each component, thereby improving the estimation accuracy of the initial frequency and the frequency modulation slope, and has a lower signal-to-noise ratio threshold. Secondly, the use of STFRFT can ensure that only one LFM component signal has the highest time–frequency resolution during single-cycle estimation, and eliminates the mutual influence between LFM component signals during instantaneous frequency estimation. In addition, when the number of LFM component signals is unknown, when estimating the optimal transformation order of the LFM component signal, by calculating the variance of the entropy vector and comparing it with the threshold, it can adaptively decide whether the parameter estimation is terminated or not, and determine the number of LFM component signals in the signal.

2 Parameter Estimation of Adaptive Multi-Component Chirp Signal

2.1 Algorithm Steps

Assume that the multi-component LFM signal consists of LFM signal components and background noise.

$$x(m) = \sum_{l=1}^{L} s_l(m) + w(m) = \sum_{l=1}^{L} A_l \exp\Big[j2\pi\Big(f_l(m\Delta t) + \frac{\mu_l}{2}(m\Delta t)^2\Big)\Big] + w(m) \tag{1}$$

where $s_l(m)$ is the lth LFM component signal, and A_l, f_l and μ_l are the signal amplitude, starting frequency and frequency modulation slope, respectively.

For multi-component chirp signals, the basic idea of parameter estimation is to estimate the chirp signal parameters of each component in turn, and after estimating the chirp signal parameters of a certain component, it needs to be reconstructed according to the estimated parameters and removed from the original echo signal. When estimating the chirp signal parameters of a certain component, the traditional time–frequency analysis methods such as STFT are used to estimate the instantaneous frequency of the interaction between the components. To solve this problem, this section proposes an adaptive multi-component linearity FM signal parameter estimation method based on STFRFT. The so-called "adaptive" means that the number of chirp signals in the echo signal is determined according to the corresponding decision rule and the loop estimation of the chirp signal parameters is automatically ended. The specific steps of the method are as follows.

Step 1: Initialize the index of the number of LFM component signals $l = 1$, residual signal $x_r(m) = x(m)$.

Step 2: Determine whether there is an LFM component signal in the remaining signal $x_r(m)$ and estimate the matching transformation order p_l corresponding to the lth LFM component signal $s_l(m)$. If there is an LFM component signal in the remaining signal $x_r(m)$, the global search method based on the minimum entropy criterion is used to estimate the optimal transformation order p_l. Otherwise, the parameter estimation ends.

Step 2.1: The transform order is discretized with the search step length Δp in the half-period interval [0, 2], and a discrete value $N = \lfloor 2/\Delta p \rfloor$ is obtained, that is $P = [p_{l1}, \cdots p_{li}, \cdots p_{lN}]^T$, where P represents the transform order candidate value vector, $p_{li} \in P$ which is called the lth transform order candidate value of the ith LFM component signal.

Step 2.2: Calculate the FrFT result of $\text{FRFT}_{p_{li}}(u)$ of the remaining signal $x_r(m)$, when the transformation order is equal to p_{li}.

Step 2.3: Calculate the entropy value $E_{p_{li}}$ of the FrFT result of the residual signal $x_r(m)$ when the transformation order is equal to p_{li}.

Step 2.4: Step the iteration number $i = i + 1$. Judge $i > N$, if it is established, go to step 2.5, otherwise return to step 2.2 to continue execution.

Step 2.5: Each candidate value of the transformation order can calculate the entropy value through Step 2.2 and Step 2.3, and the entropy value vector E corresponding to the candidate value vector P of the transformation order can be obtained as

$$E = \left[E_{p_{l1}}, \cdots E_{p_{li}}, \cdots E_{p_{lN}}\right]^T \tag{2}$$

Step 2.6: The normalized entropy value vector is E, and its formula is as follows $E_n = (-E + \max(E))\big/\max(-E + \max(E))$ (3).

where E_n is Normalized entropy vector, and $\max(\cdot)$ is maximum value operation. Calculate the variance D_E of the entropy vector E, and its formula is as follows.

$$D_E = \sum_{i=1}^{N} \left(E_{p_{li}} - \overline{E}\right)^2 \Big/ N \tag{4}$$

where $\overline{E}$ is the mean of entropy vector E, $\overline{E} = \sum_{i=1}^{N} E_{p_{li}} \Big/ N$. Determine whether the following formula is true.

$$D_E = \sum_{i=1}^{N} \left(E_{p_{li}} - \overline{E}\right)^2 \Big/ N \le \xi \tag{5}$$

where ξ is the threshold for judging the presence or absence of LFM component signal. According to the results of many experiments, ξ is set to 0.02. If Eq. (5) is established, it is determined that there is an LFM component signal in the remaining signal $x_r(m)$, and then proceed to Step 2.7. Otherwise, it is determined that there is only a noise signal in the remaining signal $x_r(m)$, and the loop is terminated. The parameter estimation is ended.

Step 2.7: Estimate the matching transformation order $\hat{p}_l$ of the lth LFM component signal $s_l(m)$.

Step 3: Estimate the signal amplitude $\hat{A}_l$ of the lth LFM component signal $s_l(m)$.

Step 4: Estimate the start frequency and frequency modulation slope of the lth LFM component signal $s_l(m)$. The rough estimate is $\hat{f}_l$ and $\hat{\mu}_l$, and the fine estimate is $\hat{f}_l^r$ and $\hat{\mu}_l^r$, respectively.

Step 5: Reconstruct the lth LFM component signal. According to the estimated value $\hat{A}_l$ of the signal amplitude of the lth LFM component signal $s_l(m)$ obtained in Step 3 and the precise estimated value of the starting frequency and the frequency modulation slope obtained in Step 4 $\hat{f}_l^r$ and $\hat{\mu}_l^r$, reconstruct $s_l(m)$ as follows.

$$\hat{s}_l(m) = \hat{A}_l \exp\left[j2\pi \left(\hat{f}_l^r (m\Delta t) + \frac{\hat{\mu}_l^r}{2} (m\Delta t)^2 \right) \right], \quad m \in \left[-M/2,\ M/2 \right) \tag{6}$$

where $\hat{s}_l(m)$ is the reconstructed the lth LFM component signal.

Step 6: Step the index of the number of LFM component signals $l = l + 1$ and update the remaining signals as

$$x_r(m) = x_r(m) - \hat{s}_l(m) \tag{7}$$

Return to Step 2 to continue execution.

2.2 Medium Frequency Time Window Function Selection

The STFRFT algorithm is used to estimate the instantaneous frequency of LFM signal, and then according to the estimated instantaneous frequency of LFM signal, the parameter estimation of LFM signal is realized by the strategy of rough estimation first and then fine estimation. Whether the STFRFT algorithm with high time-FRFD frequency resolution used in this section or the STFT algorithm used in literature, it is inevitable to select the window function length in the algorithm, because the window function length will directly affect the time–frequency resolution in the time–frequency analysis algorithm. The higher the time–frequency resolution, the more accurate the instantaneous frequency estimation, and then the higher the parameter estimation accuracy of LFM signal.

In this paper, the quasi maximum likelihood estimation (QML) method proposed in reference [2] is used to select the length of window function. The algorithm steps are as follows.

Step 1: Initialize window function length $h \in H$. Where H is the window function length candidate value vector composed of window function length candidate values h_k, and $H = [h_1, \cdots, h_k, \cdots h_K]$, K is the number of window function length candidate values. It should be pointed out that since the FRFT algorithm is implemented by fast Fourier transform to reduce the computational complexity of the algorithm, the length of the window function is an exponential power of 2.

Step 2: Calculate the quasi maximum likelihood function value corresponding to different window function lengths, and its mathematical expression is

$$J_{QML}(h_k) = \left| \sum_m x(m) \exp\left[-j2\pi \left(\hat{f}_{0,h_k}^r (m\Delta t) + \frac{\hat{\mu}_{0,h_k}^r}{2} (m\Delta t)^2 \right) \right] \right| \tag{8}$$

where $J_{QML}(h_k)$ represents the quasi maximum likelihood function value of the echo signal when the window function length is h_k, and $\hat{f}^r_{0,h_k}$ and $\hat{\mu}^r_{0,h_k}$ represent the estimated value of the starting frequency and frequency modulation slope, respectively.

Step 3: The optimal window function length $\hat{h}$ can be obtained by searching the maximum value of the quasi maximum likelihood function value, and its mathematical expression is

$$\hat{h} = \arg\max_{h_k} J_{QML}(h_k) \tag{9}$$

At the same time, the precise estimation values of the starting frequency and frequency modulation slope of the LFM signal corresponding to the optimal window function length $\hat{h}$ are the final estimation values of the starting frequency $\hat{f}^r_{0,\hat{h}}$ and frequency modulation slope $\hat{\mu}^r_{0,\hat{h}}$ of the LFM signal.

$$\hat{f}^r_0 = \hat{f}^r_{0,\hat{h}} \tag{10}$$

$$\hat{\mu}^r_0 = \hat{\mu}^r_{0,\hat{h}} \tag{11}$$

3 Simulation Experiment and Result Analysis

This section designs simulation experiments to verify the effectiveness of the method proposed in this paper and its superiority in LFM signal parameter estimation performance compared to other reference methods.

This experiment simulates an echo signal composed of two-component LFM signals, and compares and verifies the effectiveness of the method proposed in this paper for parameter estimation of multi-component LFM signals through Monte Carlo simulation experiments. The two LFM component signal parameters are set as follows: FM slope $\mu_1 = -16$ Hz/s, $\mu_2 = 10$ Hz/s, start frequency $f_1 = 28$ Hz, $f_1 = -17$ Hz, signal amplitude $A_1 = 1$, $A_2 = 1$, observation time $T_a = 2$ s, sampling points in the observation time $M = 256$, sampling time interval $\Delta t = 1/128$ s, and Monte Carlo simulation test times $N_{\text{trails}} = 1000$. Figure 1 shows the parameter estimation error of different comparison methods with the change of the signal-to-noise ratio. The comparison methods mainly include that one is the low-pass filtering process in the precise estimation process, and the other is to filter out the estimated LFM signal components and the method used to update the residual signal is different. In addition to the reconstruction and subtraction method used in this paper, there are also the STFRFT filtering method and the Dechirp processing method. It should be noted that signal 1 and signal 2 in Fig. 1 refer to the first and second LFM component signals, respectively.

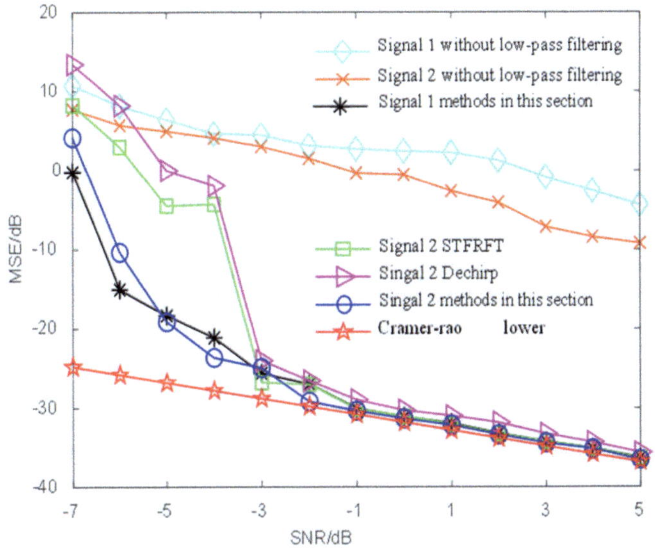

(a) FM slope estimation error

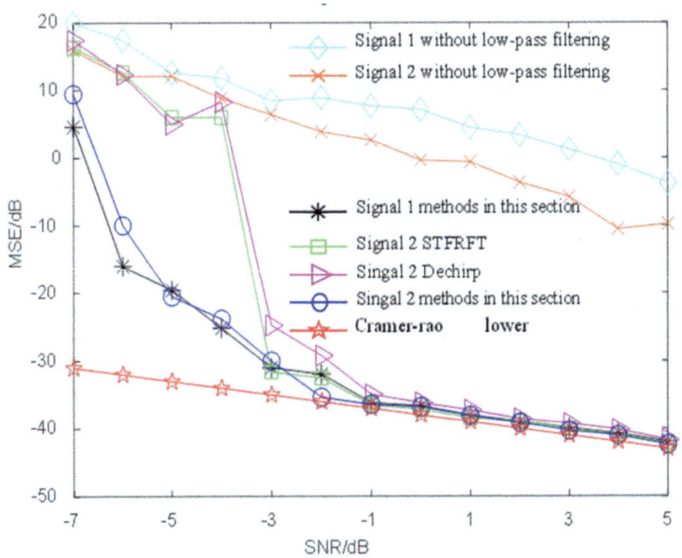

(b) Start frequency estimation error

Fig. 1 Comparison of parameter estimation errors of multi-component LFM signals with different methods

It can be seen from Fig. 1 that comparing the parameter estimation errors of the two LFM component signals with or without low-pass filtering in the method proposed in this paper, it can be found that when there is no low-pass filtering based on the moving average filter, either the FM slope or the initial frequency estimation error is obviously higher. This is because the low-pass filter processing is mainly used to reduce the echo signal-to-noise ratio, which also explains the necessity of the low-pass filter processing in the method proposed in this section. When filtering out the estimated LFM signal components, since the STFRFT filtering processing method is greatly affected by noise, and the Dechirp processing method has an error transmission problem when the rough estimation is inaccurate, the two methods have low signal-to-noise ratio. The estimation errors of the timing frequency modulation slope and the starting frequency are both higher than the methods mentioned in this paper. It can be seen from Fig. 1 that the proposed method is not affected by the number of LFM signal components, and the parameter estimation errors of the two LFM signal components are equivalent, thus verifying the effectiveness of the proposed method.

4 Conclusion

The STFRFT-based adaptive multi-component LFM signal parameter estimation algorithm proposed in this paper improves the estimation accuracy of the instantaneous frequency of each component, thereby improving the estimation accuracy of the initial frequency and the frequency modulation slope, and has a lower signal-to-noise ratio threshold. Monte Carlo simulation experiments verify the effectiveness and optimization of this algorithm.

Acknowledgements This work described in this paper was fully supported by a grant from the Guiding Project of Science and Technology research plan of Hubei Provincial Department of Education (No:B2020329).

References

1. Barbarossa, S., Scaglione, A., Giannakis, G.B.: Product high-order ambiguity function for multi-component polynomial-phase signal modeling. IEEE Trans. Signal Process. **46**(3), 691–708 (1998)
2. Djurović, I., Simeunović, M.: Review of the quasi-maximum likelihood estimator for polynomial phase signals. Digital Signal Process. **72**, 59–74 (2017)
3. Madsen, N., Cao, S.Y.: Finite difference algorithm for polynomial phase signal parameter estimation. IEEE Trans. Aerosp. Electron. Syst. **56**(1), 57–66 (2019)
4. Qu, Q., Jin, M.L.: Adaptive fractional fourier transform based chirp signal detection and parameter estimation. J. Electron. Inf. Technol. **31**(12), 2937–2940 (2009)

5. Chen, Y.L., Guo, L.H., Gong, Z.X.: The concise fractional Fourier transform and its application in detection and parameter estimation of the linear frequency-modulated signal. ACTA ACUSTICA **40**(6), 761–771 (2015)
6. Djurović, I., Stanković, L.: Quasi-maximum-likelihood estimator of polynomial phase signals. IET Signal Proc. **8**(4), 347–359 (2014)
7. Peleg, S., Friedlander, B.: The discrete polynomial phase transform. IEEE Trans. Signal Process. **43**(8), 1901–1914 (1995)
8. Swingler, D.N.: Simple approximations to the Cramer-Rao lower bound on direction of arrival for closely spaced sources. IEEE Trans. Signal Process. **41**(4), 1668–1672 (2002)
9. Porat, B., Friedlander, B.: Asymptotic statistical analysis of the high-order ambiguity function for parameter estimation of polynomial-phase signals. IEEE Trans. Inf. Theory **42**(3), 995–1001 (1996)

Author Index

L. C. Jain et al. (eds.), *Advances in Wireless Communications and Applications*,
Smart Innovation, Systems and Technologies 299,
https://doi.org/10.1007/978-981-19-2255-8

MIX
Papier aus verantwortungsvollen Quellen
Paper from responsible sources
FSC® C105338

FSC
www.fsc.org

If you have any concerns about our products,
you can contact us on
ProductSafety@springernature.com

In case Publisher is established outside the EU,
the EU authorized representative is:
Springer Nature Customer Service Center GmbH
Europaplatz 3, 69115 Heidelberg, Germany

Printed by Libri Plureos GmbH
in Hamburg, Germany